A Nurtured

A Surgeon's Life in Two Halves

Open and Closed

by

David S. Evans

ISBN: 978-0-244-35011-6

PublishNation

www.publishnation.co.uk

"Surgeons must consider the cause of the effect for without this knowledge treatment can not be instituted as it is often necessary to prevent the cause of the effect"

John Hunter FRS (1728-1793)

Acknowledgments

Without the help, advice and encouragement of others, particularly non-medical personnel, this biography would not have been completed. In particular I am grateful to Brian Evans and Judy Goodwin for reading the manuscript through and making helpful suggestions and corrections. Sir Barry Jackson kindly wrote a foreword. My greatest thanks go to my wife who supported me throughout as she did during my career as a surgeon.

Foreword

In hospital dramas and popular fiction it is almost always the case that a surgeon rather than a physician is the hero and star of the show. Tense scenes in the operating theatre and life-threatening emergencies in the accident and emergency room abound. But never depicted is the background of hard work, long training as a junior, research study and general application that with a modicum of good fortune leads to a successful career. As a consequence, members of the public have little idea as to what makes a surgeon nor what their life is really like. Neither do they know the pressures that surgeons often face when unexpected problems arise. This most readable memoir by a retired practitioner who has made signal contributions to his profession will rectify that deficiency.

The author describes his youthful years leading to medical school and qualification in medicine followed by several years of varied junior hospital appointments typical of those wishing to pursue a surgical career. He then discovers both the heartache and the joy of research. This period of his life culminates in the first of his important contributions to medicine, namely the application of portable Doppler ultrasound to the early diagnosis of post-operative deep vein thrombosis. For this work he was awarded a higher degree, gave important lectures and saw the technique become widely adopted. It has saved the lives of many patients who might otherwise have died from a pulmonary embolism. Marriage, family and a busy consultant career follow. Memorable operations are described which make for vivid reading; gradually he becomes an important pillar of the local community. The progressive evolving of surgical practice is emphasized. The early 1990s initiate another notable contribution to his art as a consequence of the introduction of minimal invasive (keyhole) surgery. One of the earliest practitioners in the UK of this new form of surgery, the author describes his extensive experience in teaching the technique and promulgating it nationally, especially in regard to hernia repair. Meticulously documenting each operation, he performs over 2500 such keyhole hernia repairs and is invited to present his experience to the Royal College of Surgeons in an important named lecture. In retirement he becomes a patient himself and experiences major surgery as well as life-threatening illness.

Such are the bare bones of one surgeon's fascinating life from childhood to mature years during which he performed over 30,000 operations. There is more, much more, but read on and be captivated. This is real life, not hospital drama or fiction, and by far the more interesting.

Sir Barry Jackson MS FRCS FRCP

Chapters

Chapter 1

My Early Years

While King George V was undertaking silver jubilee celebrations, I arrived on Planet Earth at the London Clinic on the 12th May 1935 born of parents, both surgeons.

My father, whose father was a builder, hailed from Cardiganshire where he was the oldest of four children. He decided to do medicine as a career and entered St. Bartholomew's Hospital as a medical student, qualifying in 1928. He did his first post at his alma mater and thereafter was employed by the London County Council (LCC) training as an orthopaedic surgeon with a special interest in children, at that time dealing mainly with congenital abnormalities, the ravages of poliomyelitis and bone and joint tuberculosis. He was employed from 1932 at Queen Mary's Hospital, Carshalton, where he met my mother.

My mother was born in the Scottish Highlands where her father was a factor on one of the large estates. She entered Edinburgh University as a medical student and qualified well in 1928. After registration, she went to the Princess Margaret Rose Hospital, Edinburgh, gaining a basis in orthopaedics and particularly hand surgery. While there she completed some research into tuberculosis and was awarded an MD for this work, at that time a considerable achievement for a female doctor. She was then appointed to Queen Mary's Hospital, Carshalton, to further her career in orthopaedics and especially hand surgery.

In July 1934 my parents were married at St. John's Church, Princes Street, Edinburgh, and nine months later I was the result of the union. It not being feasible for my mother to follow a career in orthopaedics, she retrained at Great Ormond Street Children's Hospital, London, attending courses to undertake child care and school clinics.

In 1938, my father was appointed Medical Superintendent at Heatherwood Hospital, Ascot, another LCC institute, and we moved there and lived in the house provided within the grounds. It was about this time that I became aware of what was happening in the big wide world. I remember having a jellyfish sting while toddling on a beach at Hastings and going for a trip on a channel steamer. Nearer home in Ascot, I recall going for a walk on the racecourse and seeing horse-drawn coaches parked by the winning post opposite the stand for the summer meeting. In the autumn I remember brushing through fallen leaves at the side of the road while out for a walk, accompanied by a pram containing my brother. The soil on farms was turned by horse-drawn ploughs. Around this time I managed to lock myself into a first-floor bathroom and, not being able to open the door, escaped out of the window sliding down the roof and falling to the ground, luckily into a flower bed. Having walked into the house through the front door I then slept for twenty-four hours. My lower spine to this day shows the marks of this episode.

We were fortunate to have one of the early fridges but as this was new summer fruits were preserved by bottling, and eggs, plentiful in summer, were placed under water-glass to preserve for winter consumption.

My first school around the beginning of the war was at North Lodge where I remember having been in a fight with the son of Lord Alexander, the World War Two general, and giving him a black eye much to the consternation of all. With the onset of the blitz on London we no longer slept in our beds but in an internal corridor within the house, being woken nightly by the gunfire from the local ack-ack batteries in Windsor Great Park. When accompanying my mother one day on the way to a clinic in Windsor, we stopped at the roadside to look at a Messerschmitt that had been shot down the previous night. On another occasion my mother was stopped by the police who had noticed that I was in the car, to ensure that she was doing an authorized journey there being petrol rationing for which she required extra petrol coupons to go about her medical duties. I was introduced to hospital life at this time going with my father round the wards at Christmas. I saw Father Christmas visit the wards and give out presents to the children, many of whom were suffering from tuberculosis or poliomyelitis. Some were breathing with the support of an iron lung. My father had come to know Captain Geary who was bandmaster of the Royal Artillery, and he on occasions brought the band to the hospital to entertain the children, a happy episode at an otherwise bleak time.

In 1942, my father was moved and promoted by the LCC to take over and run Queen Mary's, Carshalton, a return to a hospital with obvious happy memories. I went to the local preparatory school, St. Norberts, beside Carshalton Beeches station where we were grounded in the three Rs and introduced to sport. This was a pleasant time, the intrusions of the war being less than previously. The superintendent's house was on three storeys and set on the periphery of the hospital grounds with a largish garden and attached orchard containing apples, pears, plums, a mulberry tree and a variety of nut trees including walnuts. As part of the war effort, vegetables were grown and we had Rhode Island Red hens to ensure eggs. These had names and lived for some three or four years before being condemned and broiled. Despite this, although tasty, I well remember the toughness of the chew on Biddy. At leisure I could roam the nearby woods and soon met another boy, Roger. We set about building our own hide and Anderson-type shelter in the woods, excavating the ground and constructing a side wall of timber with wire to act as windows and an asbestos corrugated roof, supplies being purloined from waste sites within the hospital grounds. Towards the end of 1943, I started wandering and exploring the hospital grounds. After a while I found the hospital telephone exchange within the office block and used to help the telephonists making and accepting calls. Those were the days when to accept a call you plugged a lead into a board below a flashing light, threw a switch and spoke using headphones and attached microphone. To put through to the required destination, another plug was inserted in the appropriate hole and a handle turned for the recipient to get a ring.

The next major event while at Carshalton was the sunny evening of the 5th June 1944, when for hours planes of all sizes were flying over, heading in a southerly direction with many towing gliders behind them. Next day we learned we had been watching elements of the force beginning the invasion of France, D-day. Aerial activity overhead remained high to support the invasion, but it was not long before the Germans retaliated and started sending pilotless planes, doodle-bugs (V1s), to bomb London. I saw one flying over and as it passed the engine cut, followed by a distant bang a few minutes later. Some days later I saw another doodlebug being chased by a Spitfire. Having been through a quiet period with little overhead activity from the Germans, life now became much more tense. I was at the telephone exchange around lunch time – scromph! – there was a loud bang and windows rattled and the building shook. I had dived under a table in less than a second. Soon realizing that no gross damage had been done, I resurfaced and ran the 200 yards home to find the glass in the front door blown in, just missing my then youngest brother of four months who had been sleeping in a pram in the hall and luckily not by the door. I went with my father in the afternoon to see the shattered houses half a mile away and learned that relatives of some hospital employees had been killed. Within a few days we were evacuated, leaving my father behind to stay and run the hospital. We were transported to King's Cross station, an eerie journey in late evening and twilight passing by bombed buildings with debris still around, and put on a packed overnight train to Edinburgh. There was a major sigh of relief when Grantham was passed, the train then being out of range of the V1s.

Next morning on arrival at Edinburgh we caught another train which took us to Callander. My mother's relatives had found an old rectory which we rented and where we stayed for three months. This was a tense time for my mother because the army had been training in the hills above the rectory and using live ammo. Two months previously a cousin roaming the hills with a friend had found a grenade which exploded severely damaging a leg and requiring a below knee amputation. We were under strict instructions to keep to paths and not to pick up any objects we might find. There was a farm nearby from which we got milk and eggs, and I was allowed to go there on my own to help with the milking, this being done by hand. On other occasions we went to the banks of the nearby River Tieth with nets and caught tadpoles which, back in the rectory, we watched grow. In September, we moved to a house next to the church in

Murrayfield Avenue, Edinburgh owned jointly by my mother's brother, sister and cousin. This was our home for the next nine months.

Having run wild for three months following evacuation, in September I was sent to a school at Corstorphine, not far from the zoo. By this time the patients together with the medical staff at Queen Mary's Hospital had been evacuated to Durham where shortly after my father contracted pneumonia, a serious illness prior to the arrival of antibiotics not available for general use until somewhat later. He happily recovered but it gave us an opportunity to see him again while he convalesced. On recovery, he returned to Durham, leaving my mother and nanny to care for myself and the then three brothers, a fourth arriving the following June. I only spent one term at that school as my language and vocabulary deteriorated to an unacceptable level. While there I visited the zoo many times and soon found it profitable to act as a guide to American servicemen and their girlfriends taking them to animals and cages they wished to see. As a reward for my services more often than not I was given sweets which at that time were rationed. After Christmas I went to a preparatory school in the West End of Edinburgh, not far from the church where my parents had married and I had subsequently been christened. By early Spring of 1945, World War Two was coming to an end and we were able to return to Queen Mary's Hospital and our home. I returned to St.Norberts school for one term before moving as a weekly boarder to Kingswood House preparatory school in Epsom, a school run by Billy Malden and his wife Thelma, where I was to stay till passing Common Entrance some three and a half years later. Returning to school on Sunday evenings during the first year proved a highly emotional experience, but come Monday all was well. It was at this time that my father instilled the motto "Work hard and play hard", a scaled down version of what he had learned at St. Bartholomew's – "Whatsoever thine hand findeth to do, do it with all thy might".

In 1946, my father was appointed as lead surgeon and superintendent to the Lord Mayor Treloar Orthopaedic Hospital and College for Children with 250 beds and set in 100 acres to the west of Alton in Hampshire. This was a private hospital built and funded by donations but receiving payments for treatment from councils whose children were being treated. A band of collectors was set up early in the twentieth century by opening the then small nationwide telephone directory and approaching likely personnel. This proved highly successful. The hospital had been modernized and rebuilt from a hutted institution in the 1930s. It again treated children with bone and joint tuberculosis, those with the orthopaedic sequelae of poliomyelitis, and those requiring surgery resulting from birth abnormalities and conditions such as congenital dislocation of the hip (CDH), slipped epiphyses of the hip and those with orthopaedic injuries resulting from trauma. While the medical staff mainly resided locally, general surgery on children was done by a visiting surgeon from London, plastic surgery by a surgeon from Oxford and X-rays were reported by a visiting radiologist again from London. At the time of my father's appointment there was a large kitchen garden with some ten gardeners producing the hospital's vegetables. The hospital was overseen and managed by some five trustees including the founder's daughter, a prominent city banker, a city lawyer and the editor of one of the national newspapers, an eminent and widely experienced body. These met monthly while the day to day running was done by the steward with one typist, a hospital secretary, a matron and my father, the superintendent. How times have changed!

We lived in a house set in the grounds above the hospital with a large garden and lawn maintained by the hospital staff and with woods around.

Shortly after arriving there, the then Labour Government set about nationalizing the health service. Realizing that funds would disappear into the national pocket, the hospital was totally re-equipped with the most modern apparatus including X-ray machines which mostly came from America, British equipment being unavailable at that time due to the deprivations of the war. The college for schooling and training disabled youngsters in crafts such as tailoring and shoemaking and repairing was separated and moved to a new site a few miles distant at Lower Froyle between Alton and Farnham and remained as a separate and private entity. Despite these measures, come the day of nationalization the government took over one million pounds still in the hospital funds, enough at that time to build three new hospitals of similar size.

With the move to Hampshire and my parents' satisfaction with Kingswood House I became a full-time boarder, getting home at half term and having occasional visits from my parents at other times, usually to watch some sporting event. In this environment my sporting abilities flourished while academically my performance was just about acceptable. The winter of 1947 was cold, bleak and lengthy. One game of rugby was played before the snow arrived at the end of January. This lasted for nearly seven weeks so sport was confined to building snowmen or sliding across nearly twenty yards of ice renovated each evening in the play yard and attempting to perform on each occasion a more daring manoeuvre. Indoors, in the gym, many a game of British Bulldogs was played. That year I ran for the first time in the school cross country race with about a hundred competing and managed to come third, before winning the race in the subsequent two years. In the autumn I played in the first eleven at soccer and went on to play in the teams at cricket and rugby as well, captaining the soccer in 1948 and '49. With four younger brothers, holidays with them were spent playing cricket, tennis and soccer or having beach holidays at Easter and in the summer, first at Selsey or later Middleton where my parents had a holiday house. In the autumn of 1947 I had an injury at soccer sustaining a traumatic osteochondritis of the tibial tuberosity of my left knee (Osgood-Schlatter disease), the effects of which I have suffered from till this day. As my body grew in size, this injury resulted in a grossly misshapen knee, considerably limiting my future sporting activities. As time passed, my parents contemplated my future schooling and decided to send me to Charterhouse. I sat Common Entrance in the autumn of 1948 and entered the school at the beginning of 1949.

By 1947 I was lucky in that I had decided what I wanted to do when I grew up. My sights were set on becoming an orthopaedic surgeon. As a result of my injury I was going to the Treatment Centre at Lord Mayor Treloar's run by Sister Smith and, while talking with her when having electrical therapy, she initiated my first view of an operating theatre in action. The Treatment Centre had a flat roof. From this one could see through a window into the operating theatre below. After a few minutes my presence on the roof was observed and I had a wave from those within. On visiting the roof for the third time and thus confirming my interest in the procedures taking place, I was summoned to come down to the theatre and enter under supervision to have a closer look. I was kitted out and instructed as to where I could go and what I could touch and what I must avoid. Realizing that I was in a privileged position, I took great care not to cross any boundaries and by thus doing ensured all future visits to the environment of my adult life.

In January 1949 I entered Charterhouse with four others joining my house, Hodgsonites, at the same time. Although prepared for a change it was nevertheless quite a shock to the system having been top of the pile one month earlier as I was now starting again at the bottom of the ladder. On reflection this demonstrated one of the important lessons for life – starting at the bottom and working your way up again. During the first two weeks one was fathered and shown the ropes and learnt the jargon. For example, prefects were called “monitors”, prep was called “banco” and your preparatory school a “tutheran”. I was lucky in that being good at sport I was elevated after trials to represent my house at both the year level and under sixteens. After two weeks and passing the new boys test you became liable for fagging and duties such as cleaning a senior member of the house’s study. Inevitably one performed misdeeds for which one was punished.

Punishments varied from the most severe, a beating by the head of house to more trivial such as learning lines or being sent on a run with a senior escort to confirm that the distance had been completed in the standard time. More severe punishments, luckily none came my way, consisted of a beating by the housemaster or occasionally in extreme cases the headmaster. After a short time the only punishment which halted me in my tracks and made me peruse what I had done was a beating. If sent on a run, when I would be followed on a bicycle by someone with intermediate authority, I would delight in almost walking down a hill then sprinting up the subsequent gradient and leaving my escort behind. Only once was I put through my paces when accompanied by the school captain of cross-country running who had some insight to my latent abilities. On other occasions when set lines to learn by a monitor with poor sight, I would write them out, enter his study when empty pin and them to the board above his desk where he would sit facing me and then read them off. The monitor would then be enticed from his den whereupon I would remove the evidence. Mission accomplished! Much of my spare time at school was spent playing sport but I did enough work to creep through the necessary exams as they arose. We were fortunate in that there was much else to do. There was always a course of art lectures and everyone had to be in the Combined Cadet Force (CCF) or work in this time on the land. Baden-Powell was an old boy of the school so there was also a scout movement. During the long quarter (spring term) of 1950 there was a severe 'flu epidemic which in turn affected most of the school. I duly caught it and was admitted to Great Comp. (school sanatorium). As I was recovering six days later the lads in my ward decided to attack the next ward, which we duly did when we engaged in a pillow fight. Needless to say we were apprehended and as most of us were by now recovering we were sent back to our houses to be punished. My housemaster, one Sniffy Russell, interviewed me and asked me to describe what we were up to. I told him we crept into the next ward on tiptoe. Then what did you do, he asked. We beat them with pillows, I replied. He then summarily dismissed me without punishment. Sometime later he told me that he sent me on my way as my description of the event sent him into stitches of laughter and he couldn't contain himself.

In the holidays I was able to pursue my interest in medicine, regularly going into the theatre and scrubbing up and assisting from the age of fourteen. I learned how to tie surgical knots and soon was putting in skin sutures. And then something which would be impossible today, I did my first operation at sixteen, assisted by my father's registrar Carlo Biaggi, later a consultant orthopaedic surgeon in Scotland. I removed some staples which earlier had been put into the normal knee of a youth who had had poliomyelitis to stunt growth in the normal limb to hopefully achieve equality of limb length. At the same age I took my O levels having one hiccough necessitating a retake, and entered the sixth form as a biology student.

I studied biology and chemistry at A level and physics at O level which was a requirement to enter medical school. In the laboratory one amusing incident occurred to the boyish sense of humour when there was a shriek from the other side of the hash (class). A fellow student disappeared into the washroom after spilling hot concentrated sulphuric acid over his front. When he reappeared some minutes later, the front of his trousers had disappeared allowing all to be seen. There was luckily no physical damage but the episode although hilarious was a reminder to us all of the dangers of the substances with which we were experimenting.

It was now time to decide whither I should go on leaving school. I was the oldest of five boys, the youngest being ten years younger than me. My parents wanted us all to have equal opportunities in life so decided we should all have the same schooling which meant Kingswood House to start followed by Charterhouse, a programme with a great demand on the pocket. The two possibilities were St. Bartholomew's, my father's alma mater, and St. Thomas' Hospital medical schools. Among the visiting surgeons at Lord Mayor Treloar's was one from St. Thomas', Mr. R W Nevin, a general surgeon who subsequently became dean of the medical school. He attended once a month to do paediatric surgery. He was a man my father respected and whom I was able to assist at operations on a number of occasions when home from school. It was decided because of the tight financial situation at home that I should apply to St. Thomas' and if accepted commute on a daily basis. I applied and attended for interview. I did not interview particularly well but the committee appreciated that I came from a very medical family and that I knew what I would be letting myself in for, and that I was determined to follow in the family tradition and eventually become a surgeon. It did me no harm that Bob Nevin was a member of the interviewing committee. I was duly accepted with the proviso that if I could get physics in addition at A level I could start in the second year, otherwise I was required to enter the first year. Sadly, time did not allow me to upgrade my physics to the necessary standard so I entered the first year. The last two years at Charterhouse passed swiftly and I was subsequently glad of the all-round education I had achieved as, unlike Oxbridge or another university, one was no longer mixing with those pursuing different careers and disciplines. I much valued my time at school and although not achieving first team status mainly due to the knee injury at Kingswood House, I did captain the second eleven at cricket to their most successful season within living memory The housemaster, Sniffy Russell, regularly invited the monitors to come and enjoy a glass of draught cider (12% alc.) with him after evening prayers when he would discuss the house and our futures among other pertinent topics. All knew that I aimed to become a surgeon and shortly before we left one of my fellow leavers presented to him in front of me and unbeknown to me a medallion with an inscription reading "Please not Evans". He thought this highly amusing and wore it round his neck for the remainder of his life. I duly achieved the necessary exam results and left Charterhouse in the summer of 1953.

Chapter 2

Medical School

On leaving school in July 1953, I went to Middleton near Bognor Regis and stayed with a widow, whose husband was Captain of HMS *Hood* when it was sunk during the dark days of the war, and her son Graham. The reason for being there was to do a job to boost my bank balance for a holiday the following year. I cycled each morning into Bognor to work as a deckchair attendant on the sea front. At that time Londoners took day trips by charabanc (coach) to the seaside for their holidays. The deckchair stacks would be manned by a retiree and two students. The trippers would hire a chair paying in addition a deposit refunded on return of the chair. In the morning we dismantled the stack and then reassembled it on return of the chairs in the evening. As our good deeds for the day, we would carry chairs on to the beach for the elderly or infirm. I was lucky in that I and another student were teamed in the main with a retired publican from London, who opened our eyes to the big wide world and taught us among other things the rudiments of betting. After a month my parents who owned the adjacent house, to where I was staying, arrived down on holiday with my brothers and I moved back with them. On one occasion when returning home at late dusk a policeman was cycling in the other direction and gave chase to me for having no lights. With a surge with what I later learned was adrenaline I managed to outrun him and turning a corner out of sight in the private estate disappeared into a garden hiding the bike behind a bush and watched for ten or more minutes, my heart pounding while the copper searched up and down for me. Failing to find me he eventually departed leaving me somewhat chastened to complete my journey home. After eight weeks and with what in those days seemed a healthy bank balance, I returned home and joined up with a great friend in Alton, Morton Wilson, who was starting medicine at Guy's Hospital at the same time as I was entering medical school. Morton's father was a GP in Alton. Morton and I had many interests in common including golf and would meet and discuss life late into the night. Our friendship continued and some years later we became godparents to one of each other's children.

After the lengthy vacation, and opting to train as a doctor prior to being considered for National Service, I entered St. Thomas' Hospital medical school in October 1953. I did a daily commute from Alton to Waterloo on British Railways, leaving home shortly after seven and driving the two miles to the station in a 1937 Hillman Minx given to myself and my brothers by our Scottish uncle. During its early life it had been laid up for most of the war and almost certainly had not been driven at more than 40 mph. It proved ideal for the short daily journey. On arrival at Waterloo I did the fifteen-minute walk to the medical school usually arriving in time (British Rail permitting) for the first lecture. The first year proved a good opportunity to find my feet at the medical school and in London as I only had to take physics at the end, having previously gained exemption in biology and chemistry at A level. There were some thirty students in this year of which about five had done National Service prior to joining the medical school. These were much more men of the world and a good influence on the rest of us who had just left school. The quiet first year enabled me to adapt to the new life and the travel involved. St. Thomas' had excellent sports facilities at Cobham so I played soccer and the occasional game of rugby in the winter and cricket in the summer. During the winter after sport and a clean-up in the communal bath, there was always a noisy atmosphere in the clubhouse where the afternoon's activities were relived prior to the rendering of well-known bawdy rugby songs. To get home from Cobham involved getting a bus to Weybridge station to catch the train back to Alton. The wait on Weybridge station would always be enlivened by express trains roaring through drawn by steam locomotives emitting red-hot embers into the night sky as they thundered past. In the summer evenings on returning home I would often have a game of tennis with my father. At the end of June I duly passed the physics exam and set about enjoying the three-month holiday.

That summer I set off with Morton and two other friends for Spain. I travelled with Morton in a Standard 8 saloon while the others were in an open top MG. It was not long after the Spanish Civil War and a visa was required for entry. We went just beyond St. Sebastian to Zarautz, a village on the coast. This being for all of us our first trip into Europe, we revelled throughout our first morning in playing football in the sun on the near deserted beach. Ignorant as we all were of "Mad Dogs and Englishmen", we all suffered sunstroke dampening our activities for the next two days. After a few days exploring the area we then crossed Spain to the Mediterranean coast ending up at Tossa de Mar. This was a small fishing village unspoilt and undeveloped. On going into the bar we learned we were the first English to visit since the war. Liqueurs were as cheap as water and the locals did not allow us to buy any drinks throughout our stay. Barcelona being nearby, we witnessed our first bullfight. After an enjoyable trip we returned to England to enter in October our second year as medical students. In the meantime I continued to watch and assist at operations, hip replacements now being available and done. I travelled with my father on occasions to Queen Elizabeth's College for the Disabled of which he was chairman, in addition to being visiting orthopaedic surgeon. There he examined and assessed newcomers in addition to other trainees as necessary. It was here that one marvelled at the attitude of even the most severely disabled who almost universally would say that they could cope as there would always be others worse off than themselves. Seeing these unfortunate people and how they strove to cope made one realize how fortunate one was oneself.

In October, I returned to St. Thomas' to begin my second year of medical studies. We had lost one or two students but now numbered fifty, the year being enlarged by those with first year exemption. Our studies consisted of anatomy, physiology and biochemistry when we were fortunate to have as teachers among others Professor Dai Davies, head of the anatomy department and editor of *Gray's Anatomy* and Professor Henry Barcroft FRS, head of physiology. Whereas for me the first year had been leisurely, we now entered the slog of acquiring knowledge in the basic sciences. I commuted five days a week and had lectures in all subjects combined with practicals, doing cadaveric dissections, biochemical experiments and physiological studies. We had anatomical viva voces (vivas) every three weeks and exams at the end of the terms. These studies I found tedious but realized that they were a necessary hurdle to cross prior to entry to the clinical course. I continued to play winter sports and cricket in the summer. I failed to identify how much work was required and was unsuccessful in anatomy at the end of the year, needing to retake at the beginning of the third year.

After the second year I hitch-hiked with an old school friend, Alan Traill, who later became Lord Mayor of London, down to Rome and back. We were lucky in that we were able to arrange a lift to Monte Carlo with the catering officer at Lord Mayor Treloars' Hospital who was going there on holiday and had room for two in his car. My colleague's parents had friends there and we were able to stay with them until we moved on. We soon found that we could swim from the public beach to the private areas where we could enjoy the far superior facilities. On the fourth day we set out with our backpacks and journeyed down to Rome, stopping on the way at Pisa to climb the leaning tower. In 1955 folks were very generous in offering lifts, the only difficulties being in border areas. The hitches were very variable ranging from lorry cabs to a Ferrari saloon the owner of which was pleased to demonstrate its abilities and performance. We were able to communicate mostly with primitive English or schoolboy French. In Rome we had a pleasant four days staying near the station and visiting the sights, usually setting out early to avoid crowds, in particular getting to the Sistine Chapel when very few were there. We managed to get tickets to the Baths of Caracalla, home to Rome Opera in the summer, and see an opera with Giulio Neri and Tito Gobbi as lead singers. Before leaving as tradition demanded, we threw three coins into the fountain (of Trevi), and happily have both since returned on a number of occasions to this beautiful city.

On leaving Rome somewhat late in the day, we made slow progress still hitching after dark. After a while a lorry went past and stopped fifty metres up the road and turned a search light on us. This proved to be an Italian army vehicle which gave us a lift for twenty-plus miles before dropping us just short of their destination where we spent a cold night sleeping on the ground in an orchard. On the way north we visited Siena and then Florence where we spent three days visiting art galleries and other treasures. While there we had our first experience of mass production when visiting the cathedral on a Sunday afternoon. About forty families were having their young christened. They queued up before the priest, each one announcing the name for their child who was then baptized, and at the utterance of Amen was moved on to make way for the next in line, the whole process taking less than a minute. From Florence we went to Verona, spending the first night in a dormitory at a monastery where we were eaten alive. The next evening we went to the amphitheatre and saw *Aida*, the lead singers again being Giulio Neri and Tito Gobbi. To this day some fifty years later, I still have a vivid memory of the Triumphal Entry featuring camels and other large quadrupeds with the chorus ringing out round the theatre with near perfect acoustics. That night we slept somewhat uncomfortably on a bench at Verona station where at least we were not food for hungry insects.

Next day we continued our hitch-hiking towards Venice. About eighty miles short, we were picked up by a businessman who took us into the city and let us stay for the duration in his flat near the centre, thereby much lessening the burden on our pockets. While there we journeyed by boat to Murano to see glass production. It was soon apparent where we alighted at the waterfront that the shops had products vastly more expensive than the premises further ashore where they were made. I bought six brandy balloons with wonderful engravings which I hitched back with me and still have to this day. After four days in Venice we said goodbye to our host and returned to England via Innsbruck, Munich and Aachen, travelling through Germany separately to speed our return. The journey through Germany was memorable for the quality of the autobahns, motorways being unknown in England at that time, and the destruction of towns by war particularly Aachen where there was still much rubble and buildings pockmarked with bullet holes.

The third year studies began in October with my having to retake anatomy. By now second MB was on the horizon, the exam being the following March, and I was aware that if I was to fulfil my desire to become a surgeon some serious study was required. My performance during this term improved but not sufficiently to be reassured that all would be well. It was decided that during the Easter term I would have digs in London to enable greater time for study and revision or, as an anatomist put it some years later, learning it for the first time. During the Christmas break I studied in my father's office in the hospital working mostly six to ten hours a day. Returning to London for the Easter term I resided in a flat in Prince of Wales Drive, Battersea, in the mansions of that name. My landlady was sympathetic and encouraging and I was able to work from nine o'clock in the morning, first in the hospital spending much time in the library and then in my digs when I would work through till two o'clock the following morning. This pace I adhered to for six weeks by which time I was getting exhausted and stale, not being able to remember something I had read five minutes earlier. I had to take a break and went home with considerable trepidation and without a book or notes to study. I slept long hours and played two rounds of golf. On my return to London four days later, I was surprised to find how refreshed I was and that I could then remember in detail what I had studied some weeks before. I resumed the same work pattern which I kept to until four days before the exam when, again exhausted, I went home to recharge the batteries. With plenty of rest and again a round of golf the therapy was once more successful. On returning to London I found I was rested and quietly confident that I had made up for my shortcomings in my earlier periods of study. Always thereafter I was to take three or four days holiday prior to an exam, a system that never let me down.

We were shocked and stunned on the Monday morning when arriving for our first exam to learn that one of our year, who prior to entering medical school had done National Service, had committed suicide two days before. It transpired that his relationship with his girlfriend had ended some days earlier. We all pushed for a postponement of the start of the exam but the authorities in their wisdom, and I believe rightly, decided that we should continue. Life must go on. After three days of papers, practicals and vivas we all assembled in the early evening at the bar of St. Thomas' House, the students club. We were joined by Professor Dai Davies, as Welsh as they come, who read out our results in alphabetical order. The twenty percent not successful quietly slipped away while I was one of the lucky ones who stayed to celebrate, the events of that evening rapidly disappearing into a haze beyond recall. After forty-eight hours I returned home and saw much of Morton, who had also been successful at Guy's, when we chatted late into the evenings and spent many days on the golf course. Two weeks later I recommenced the commute as we began our clinical course. By now I was joined on this journey by my brother Peter in articles and training to be an accountant. Each morning we would set off invariably late from the house to catch our train and having to disregard the speed limit through the town. One of us would be on cop watch while the other drove. Only once did we miss the train when the police were waiting for us and we had to drive circumspectly with a tail behind. The train was just pulling out of the station as we arrived.

Commencing the clinical work in April 1956 was like moving from winter to summer overnight. There were some introductory lectures and then we attended casualty where we soon were able to assist in the treatment of patients. With my earlier experience of stitching and my enthusiasm to perform, I was soon able to suture and dress wounds or drain abscesses after freezing the skin, albeit under supervision. The first six months were quiet the only course of lectures being in pharmacology, and as there were some forty of us to experience work in casualty this gave a considerable amount of free time which enabled visits to the Oval to watch cricket or to play golf with others in my year. We were joined on occasions by a senior student, John Ormsby-Gore, whose brother later became British ambassador in Washington. John was supported by his grandmother who had undertaken to fund him while he remained a student. He was already a clinical student when I entered St. Thomas' and continued as one well after I qualified. He was a most likeable person and a great participant and overseer of student activities. Newly arrived from school, after any parties at which we might have overindulged he would ensure we got home safely. He was a sort of father figure to the young and asset to the students club. The workload in the first six months being light, and Alan having finished in Cambridge, we arranged a further holiday together. We decided to visit Scandinavia on a motorcycle. While both of us had driving licences for cars, neither of us had those required for cycles. Alan had friends in Dublin where you did not have to pass a test to get a licence. Licences duly arrived and we set off from London to Harwich on a BSA Bantam where we caught the overnight ferry to Esbjerg. It was a gorgeous evening and night with a flat, calm sea. We soon joined a deck party and joined in dancing with a Danish group who we later learned were returning to their native country after performing at the National Eisteddfod in Wales. On landing we crossed Denmark to Copenhagen where we spent a few days during which we visited the Tivoli gardens and the Carlsberg brewery and saw a production of *Hamlet* in the open air at a castle. We then crossed by ferry to Helsingborg and set off to get to Stockholm. We rapidly realized that the Bantam was underpowered to carry the two of us plus luggage on this lengthy journey. We alternated one travelling on the bike with the luggage and the other hitch-hiking, having arranged to meet up again at a previously determined railway station along the route. After a few days enjoying the delights of Stockholm we returned to England for me to return to my medical studies and Alan to become a Lloyds underwriter.

After the six months introduction to the clinical course, our training began in earnest. I paired up with Tony Wing who had done his preclinical course at Oxford before arriving at St. Thomas' to do his clinical studies. During the year we did three months medicine then two lots of surgery before returning to medicine for a further three months. We were expected to know about each other's patients which we had been allocated to clerk and observe while in hospital. Each firm had eight students attached. During this time we received lectures, clerked patients and attended ward rounds most of which were teaching. It was during this time that I encountered death for the first time. My patient, whom I had clerked in, had renal failure. This was in the days prior to the development of renal dialysis and therefore meaningful treatment. I had visited him each day to monitor his progress. Inevitably, one chatted with him but instead of improving he gradually deteriorated until one morning he was in a coma, and while I listened to his heart the beat suddenly stopped. This proved an emotional experience, the memory of which has lived with me to this day.

While doing surgery we scrubbed up to assist with the operations on our own patients. I was ever ready to take another student's place should he be absent and lived in the theatre whenever my firm was operating.

Following this year we moved on to do midwifery and gynaecology, where again we were designated patients to clerk. Doing Midda, two memories stick out. The first was of an attractive auburn-haired lady who had trained as a nurse at Guy's and was married to a stockbroker. She was admitted to hospital in labour and remained in labour for ninety hours before delivering a healthy child. By this time the mother and team were exhausted. Almost certainly today she would have had a Caesarean section many hours before. The other concerned a home delivery while on district. Mothers-to-be if deemed fit and suitable for a home delivery were instructed to ring the hospital when they went into labour whereupon we would set out with a midwife to supervise the delivery. It was at night but I remember looking up at the block of flats and seeing an agitated man pacing up and down on a balcony, cigarette glowing in his mouth. There was no doubt as to where we had to go. Again, all was successful.

During my final year I lived in London at 79 Lambeth Palace Road next to the Two Sawyers pub. Between this and St. Thomas' House, the students club and residence, was one of the main nursing homes, Riddell House. Opposite on the other side of the busy road was College House where the resident medical staff lived. Of my three companions during this time one became a consultant dermatologist and another a consultant physician. In this year we studied among other disciplines orthopaedics, ENT and ophthalmology. It was now the knowledge I had gained from Lord Mayor Treloar's came to the fore and at that time I had definite visions of becoming an orthopaedic surgeon. During this part of the course I learned how long fractures would take to heal and be sound, and saw again many unusual orthopaedic conditions which previously I had encountered at Treloar's or Queen Elizabeth's College for the Disabled. When attending an outpatient clinic run by Harold Ridley we both noticed that we were wearing the same old school tie. At the end of the clinic he invited me to stay and chat. He soon told me of advances he had made and the frustration he had in getting his work recognized by his peers. It transpired that during the war he was removing Perspex from injured pilots' eyes in particular during the Battle of Britain. While doing this he noticed that the foreign body often having been there for some days had not induced any reaction in the eye. That is to say it was inert. Shortly after the war when removing a cataract from an eye which involved subsequently wearing thick ugly heavy glasses to have any vision, a student asked why, if he could remove a lens from the eye, he could then not replace it. He remembered the pilots of some years previously and set about developing an implant through connections he had in Brighton. After three successful operations he decided to publish his results and read a paper at an ophthalmological meeting. He was scorned and ridiculed by the society under the belief that the implant would either cause infection or be rejected as a foreign body, the then current thinking relating to transplantation. Some years later Harold was made a Fellow of the Royal Society and on his ninetieth birthday the Queen made him a Knight of the Realm. Today, several million patients a year throughout the world undergo successful cataract operations with the restoration of sight as a result of this chance remark and the thought it provoked.

On 5th November, Guy Fawkes' night, Lambeth Palace Road became a battleground with rockets being fired in both directions above the traffic including double-decker buses from our residence to the flat roof outside College House. Surprisingly there was no damage and only one injury with a burn to the hand. Studies were now serious with the final exams looming in April. During the days I attended outpatients and ward rounds clerking and presenting patients whenever possible, and then working well into the evening ensuring that there were no major gaps in my knowledge for the tests ahead. Once more I took time off before the exam commenced and returned to London refreshed and quietly confident. The written work was done in three days but it was three weeks before we had all completed our clinicals and vivas. Once more we gathered in St. Thomas' House for the results to be published. On this occasion in May 1959 there were few failures and celebrations went on well into the night for what was to be the last time that our intake would all meet up together.

Two days later, before starting work, I returned home for a break when again I met up with Morton, also successful, whereupon we had a month off. The need for me to work was apparent as my father, with my four younger brothers to support, said, “now you have the qualifications you can stand on your own two feet!”

Chapter 3

House Jobs

During this month we reminisced about our student days and looked to the future and what it might hold for each of us. Morton intended to become a general practitioner following his father into a practice in Alton, while I intended to become an orthopaedic surgeon. I started applying for locum posts to fill the time until pre-registration house jobs commenced in August. On one day I played cricket for St. Thomas' Hospital in a knockout cup match. The opposition batted first and then I opened the batting for our side against a respectable total. We won the match by six wickets when I was ninety-three not out. To perform well with the bat I needed to be in practice so decided there and then to retire from the game on a high, the opportunities to play in the future appearing to be very limited.

Towards the end of May I received a phone call from St. Thomas' saying that a surgical locum post due to illness was available at the country branch at Hydestile, near Godalming. During the bad years of World War Two this hutted hospital had seen much activity, patients being transferred there for longer term medical conditions and elective surgery. It now served as a centre for minor and intermediate surgery and longer term medical care patients such as those who had suffered a heart attack and who at that time were kept in hospital for one month or more following the incident to recuperate. There were resident medical and surgical officers both well up the training ladder to becoming consultants, and visits would be made by consultants from London to oversee the care and wellbeing of the patients.

It was a relaxed and undemanding way into clinical practice with time to play tennis and enjoy the company of both the medical and nursing staff who were rotated during their training from London to do a six-month spell in the country. My monthly salary while doing this job was £19 all found. At that time it would have taken more than five months earnings to buy a return flight to Barcelona. How times have changed.

At the beginning of July, wishing to have a general surgical training prior to specializing in orthopaedics, I heard that I had been appointed to St. Thomas' hospital to do my pre-registration year first as a casualty officer for six months followed by a house surgical job attached to the Boggon/Nevin general surgical firm. Each surgical firm consisted of a senior and a junior consultant, a senior registrar nearing the end of his training, a registrar and a house officer. This appointment gave me great pleasure as it set me on the path for a surgical career, which would have been challenging without a teaching hospital appointment. I duly finished at Hydestile and, a few days later, started in London.

The resident medical staff had a sitting room and their own dining room in College House at the centre of the hospital, but during the first six months the casualty officers had rooms on a floor in St. Thomas' House across the road from the hospital. There were ten new casualty officers appointed every six months, the five later doing medical jobs spending more time in the surgical side of casualty, while the likes of myself worked more in the medical rooms.

Patients on arrival in casualty would first be sorted by sister casualty or one of her charge nurses (qualified nurses proudly wearing their Nightingale badges). Peggy Woodhead, the sister, was a real matriarch, highly efficient with an easily audible staccato voice, rarely wrong in her assessment of what came through the doors and much admired by most. She ran what today would be called a five-star unit, and any inefficiency was not tolerated. Everyone knew where they stood, and it was a happy place to work. The patients were in the main local Lambethians, visitors to London, parliamentarians from across the river, emergencies from Waterloo station and police officers, St. Thomas' being the London police hospital.

The house officers were supervised by a senior medical and a senior surgical officer doing their third jobs. Above them there was the Resident Assistant Physician (RAP) and Resident Assistant

Surgical (RAS) officers, both towards the completion of their training, looking and waiting for Consultant posts. Once patients had crossed the door and been sent to the appropriate room they were clerked by the Casualty officer or in hours when present by medical students. Simple procedures like the drainage of an abscess or syringing ears were done without referral. Anything more complicated, particularly at the beginning of the job until one's abilities, had been assessed, was referred up the line for confirmation of any action suggested. Any potential admission to hospital required assessment by the RAP or RAS.

We worked twelve days at a time rotating from department to department. This enabled some periods off and I well remember going to the Oval one afternoon and watching Surrey playing Derbyshire in a county match. Surrey were batting and soon Peter May, an Old Carthusian, and Captain of England, was at the crease and scoring prolifically. Derby had two fast medium bowlers, Gladwin and Jackson, on the edge of the England side. May in one over hit a slightly short ball from one for six over extra cover and, in the next, the other off the front foot for a slightly over pitched ball again for six over mid off. In this rampant mood he went on to make a 100.

On alternate weekends when working, in addition to Casualty, I covered and cared for patients on the Boggon/Nevin firm while the House Officer had forty-eight hours away. It was about this time that the emergency call system was upgraded from coloured lights to bleeps. This resulted in one always being contactable while on hospital territory, even when taking a comfort break.

Not having done National Service prior to entering Medical School, I was summoned during this six months for an interview and Medical examination prior to being called up on completion of my medical registration. This was to be the last entry before National Service ceased. The interview was fine but when it came to my medical, my knee previously injured at Preparatory School, and now deformed, caused me to be downgraded so that only in time of war would I have been accepted. However, in peace time because of the risk of further injury and the possibility of then claiming a pension for life I was turned down. I was aware that if I had done Service I would be competing for the next surgical training posts with others who had not, and that in two years one's face could easily have been forgotten. Although missing out on the possibility of seeing some of the world, courtesy of Her Majesty, nevertheless in the long term it probably benefited my surgical career.

After six months the Casualty post came to an end and following two weeks holiday I became House Surgeon on the Boggon/Nevin firm. The firm had thirty beds for elective surgery and, in addition, others to which patients were admitted as emergencies under our care during our week on duty. This was a physically demanding job. I would meet Mr. Nevin at 7.45am in St.Thomas' Home, the private wing of the hospital, and review his patients therein. If necessary we would then go to see any patients causing concern in the main hospital. I would escort Mr. Nevin to the Dean's office before having breakfast in College House. At 9 o'clock I would meet the Registrar or Senior Registrar to do a round of the patients prior to doing a teaching round or attending outpatients. My duties included liaising with Theatre Staff to notify them of operating lists which began at 1.30pm. I would attend theatre and, for major operations, would assist together with the Senior Registrar or Registrar and a medical student. Towards the end of the list I would assist the trainees in lesser procedures.

At that time, there being no intensive care or recovery wards, patients were returned to the surgical wards in an often far from conscious state to extrude the ether anaesthetic from their systems. This after a major lengthy operation could take forty-eight hours before the patient was fully compos mentis. It was at this time that some anaesthetists were beginning to use paralysing drugs which meant that patients were paralysed and lightly asleep, as opposed to being taken down to the depths of anaesthesia, thus enabling speedier recovery and lowering the risks associated with prolonged recovery. These included pneumonia, and deep vein thrombosis which could result in pulmonary embolism and death. To prevent these, active treatment was required by the nursing staff and physiotherapists and diligent attention by all, ensuring patients were turned regularly and pressure points treated to prevent bedsores.

Patients being treated at St. Thomas' were fortunate in that the nursing standards having been set by Florence Nightingale and maintained by her successors were second to none. Woe betide any nurse who allowed the semblance of the beginning of a bedsore. From this, it will be apparent that, following operation lists, the house officer would need to check the recovering patients at regular intervals for the next few hours. This meant on a quiet day doing a night round beginning at about ten thirty by which time the night staff would have been briefed and have done a round of the patients to ensure all was well. On this round, the patients would again be checked and for those going home the next day discharge summaries would be completed and letters written to update their GPs on their progress. If all was quiet these rounds could be quite social, chatting with the nursing staff and enjoying a night cap. I well remember that on one occasion my drink was spiked with a laxative and I spent most of the night trotting. Needless to say working the long hours that we did, on this occasion I was not amused. The night round would be finished anytime between one and two thirty, enabling a few hours sleep before again meeting Mr Nevin in the morning.

One week in four the firm was on for emergencies (Major week or Take). Any general surgical emergency requiring admission would be admitted under the firm doing major week to either City (male) or Alexandra (female) wards. Common emergencies included appendicitis, strangulated hernias, perforated stomach ulcers, large bowel obstructions and injuries. After the formalities of admission, patients requiring surgery for these conditions would be taken to Basement theatre for operation, emergency lists more often beginning at nine in the evening when the night staff came on duty. On busy days, these lists could go on for three to four hours. The house surgeon would assist the RAS with these, together with any students also on major week. On completion of the operations the house surgeon would then do his routine night round checking patients and doing the necessary paper work before retiring to bed anywhere between three and six am., again rising to meet the consultant at 7.45. To keep going I would have a hearty breakfast, lunch, afternoon tea in College house, often with a consultant and registrar, dinner and then a further meal in the nurses dining room at midnight. During the six months I put on two stone in weight and came to resemble Mr. Nevin who was somewhat rounded so that we looked like Tweedle Dum and a somewhat younger Tweedle Dee. However I found with these increased energy supplies I could keep going, often with a thirty minute catnap after dinner, and manage the sleep deprivation although on weekends off I spent most of the time away with the fairies. On reflection looking back over the years I realized that during this time I gained an enormous amount of experience and learned to recognize problems early and set treatments in motion, thus nipping potential disasters and complications in the bud. I much benefited from advice given by my father who said the ward sisters in most cases with many years of experience behind them would almost certainly have seen problems before and could advise on action required. This very much proved true. If uncertain as to what was happening there was always a senior medical resident to consult and from whom to learn. During my subsequent career I was ever grateful that I had seen problems first hand as they never appeared the same when reading from a book or hearing about them in a lecture or symposium. Over the years countless patients benefited from this learning and I am certain many lives were saved.

During this six months there was a considerable upheaval as Mr. Boggon the senior consultant became ill and had to take early retirement. He was replaced by Colin Graham-Stewart, a senior registrar awaiting a consultant appointment, as a locum before the substantive appointment of Mr. Lyn Lockart-Mummery as the new consultant. Both these surgeons were sound clinicians and good technicians and I learned a lot from assisting. With the benefit of having been in operating theatres and assisting at operations from a young age, I began, admittedly by putting myself forward, to operate under supervision. With good instruction and patient senior trainers my case mix and repertoire expanded at a good pace.

Towards the end of the six months I applied for and was appointed for the following six months as the Senior Surgical Casualty Officer. I looked back at the end of my first year after qualification with satisfaction: satisfaction that I had been appointed to St. Thomas' one of the leading teaching hospitals in the country, satisfaction in that I had learned a lot in organizational and people management skills and satisfaction that one had survived the considerable physical demands and, at the same time, learned much that was to the benefit of patients in the many years to come.

After joining the family in Sitges near Barcelona, a holiday that absorbed most of the previous six months earnings, I returned to St. Thomas' as a fully registered doctor to take up my new post.

As Senior Surgical Casualty officer (SSCO) I had moved one step up the ladder and was now overseeing the newly qualified doctors doing their first job. These were going mainly to do medical posts for their second six months. After initial examination by the Casualty officer, I would check any patients requiring surgical admission to hospital prior to referring to the RAS to authorize admission. At times, when multiple accidents crossed the threshold, it would be my responsibility together with Casualty sister to triage the patients and ensure that they were seen and examined in the correct order to ensure maximal survival. With multiple major accidents the RAS would be summoned to supervise.

The RAS at this time was Bill Bradfield who had been a major in the tank regiment during WW2, and had trained post-war to be a surgeon. During his student days, he had been a hooker in the hospital first rugby fifteen when the hospital had won the inter hospitals cup. This was a much sought after trophy and more often than not won by St. Mary's with their legions of Welsh players, including JPR Williams the Welsh fullback and later a trauma orthopaedic surgeon. It was always said that the Welsh went to St. Mary's, because the adjacent Euston station was the gateway to Wales. Bill was splendid to work with and advanced my surgical skills rapidly teaching and mentoring me through many procedures. Unless very urgent, in which case patients would go to theatre as soon as possible, emergencies would be admitted during the day and taken to the Basement theatre for an operation list starting after nine in the evening when the night staff came on. At first I assisted Bill with operations but soon, he assisted me before he left me to operate once he recognized that I would call him if floundering or in difficulty. It was thus that I was able to clock up over eighty operations for strangulated hernia during six months, a figure at that time most unusual but today totally unachievable by trainees with that amount of experience. Even surgeons at the completion of their training and seeking consultant appointments on occasions can fall short of that figure.

It was not all work and no play. Many senior nurses shared flats in St. Thomas' Mansions, now no more, adjacent to Westminster Bridge road and across the road from the hospital but still well within bleeper range. Social gatherings would occur after duty giving on site relaxation. I became friendly with a Charge Nurse in Basement theatre, a member of a shipping family, and we had a lovely day out one Sunday in North Surrey. We departed that evening looking forward to further outings. Four days later, I learned that in the interim she had become engaged to a Thomas' doctor two years my senior who had returned from the USA. Such is life and it proved the end of a flourishing friendship.

On another weekend, I attended the wedding in Kent of one of the Casualty Charge Nurses who was marrying a submariner, a still memorable occasion. I well remember the best man and a fellow naval officer reading a telegram, a traditional function at weddings at that time, which stated "Congratulations to you both. Report depth and position at midnight".

It was not all plain sailing, however, and there were ups and downs. That Christmas there was an accident on Westminster Bridge when a Rolls Royce crashed. A young attractive woman in her early twenties had banged her head on the inside of the car. On arrival in Casualty she was compos mentis but soon lapsed into unconsciousness. An intra cranial haemorrhage was diagnosed and she was transferred to a neuro surgical centre in London but sadly did not survive.

Bill Bradfield and I were on duty on Christmas Day. One of my duties that day was to carve a turkey on Nuffield ward where I had cared for Boggon/Nevin male patients in the previous six months. At that time, patients stayed in hospital for anything up to two weeks or more after operation. Patients, especially if lonely, would be happy to be admitted to hospital for operations over the festive season. Carols would be sung and patients would have a present off the tree. Great care was taken to ensure that those not so well were properly looked after in the true Nightingale tradition, no corners being cut in their required care.

All was quiet on the admission front until after four in the afternoon when suddenly patients started streaming through the Casualty doors. Many had put off their journey to hospital until after the family celebrations in some cases for as much as three days. As a result, the condition of some patients was far worse than it might have been. We operated from nine o'clock on Christmas night to nearly five in the morning, some six or seven patients being treated in that time. Thankfully we were relieved at midday and joined our families, Bill kindly giving me a lift home.

It was now time to consider the next step in my training. To become a surgeon it was necessary to become a Fellow of the Royal College of Surgeons. There were two hurdles to cross, the first the Primary Examination consisted of passing advanced exams in Anatomy, Physiology and Pathology. The second at a later date was a test in theoretical knowledge, examination of patients with questions regarding incisions and a specific operation, a test of knowledge of surgical instruments and finally a further test on surgical pathology.

I applied for and was appointed an Anatomy Demonstrator in the St. Thomas' Hospital Medical School. I finished the SSCO post at the end of February and after two weeks holiday embarked on the next step of my career.

Chapter 4

The Surgical Fellowship

On returning to London, I found digs over an off-licence in the Latchmere Road, Battersea which gave me an easy bus ride into the Hospital and Medical School in Lambeth.

The anatomy department was run by Professor Dai Davies, an energetic man, who among other positions was editor of the 33rd and 34th editions of *Gray's Anatomy*, the encyclopaedia for all things anatomical. He was a proud and very Welsh Welshman who was also a Lord Lieutenant in his country, an honour and duties of which he took seriously. I and my fellow demonstrator would often work in the anatomy department well into the evening. Professor Davies himself often worked late but would always have a word with us on leaving, wishing us well and advising us that he wanted no hanky panky. Being innocents we never really discovered what he was getting at!

The demonstrator posts previously not funded were now salaried giving us sufficient to keep body and soul together. We would help the medical students with their dissections and display the anatomy as required on the cadavers thus teaching them and further familiarizing ourselves with anatomical layout. Usually, I would work in the anatomy department during the day and thereafter retire to the library to further my knowledge of physiology and pathology. The library closed at nine o'clock, whereupon back to my digs to continue study stimulated by multiple cups of coffee into the early hours of the morning. The hurdle to be crossed was challenging, only thirty percent of entrants passing Primary at first attempt. I resolved therefore to keep my head down and to temporarily ostracize myself from society and the outside world.

To enhance the prospects of passing I joined tuition classes held in Kensington Church Hall by Mr. Stansfield the anatomist at that time at the Royal College of Surgeons. There were about twenty five of us from the UK and Commonwealth, all eager to absorb his words of wisdom. He was a well travelled man having been in his time to Sydney and who also had a photographic memory for street layouts. He astounded the Australians by his knowledge of the geography and topography of their city. His method of teaching anatomy was novel. He would draw diagrams which we would copy. He demonstrated that we did not need to remember anatomical detail, so when answering a question we would draw the relevant diagram. From this we could read off the anatomical relationships to get the answer to the question. This simplified learning and left space in the memory bank for other knowledge. His sessions combined with humour and knowledge were brilliant giving us an excellent insight of the anatomy we needed to know as well as some quirky questions that might be asked during the oral part of the exam

During this period my father invited me to accompany him to a meeting of the Orthopaedic Section at the Royal Society of Medicine. He had previously been President of this section (1956-57), and thought I would be interested to hear a lecture given by Sir John Charnley FRS of Wrightington Hospital who pioneered hip replacement in the UK. The lecture was on hip replacement using the prosthesis bearing his name. The first third of the lecture consisted of the physics involved to achieve a good result. Coefficients of friction, lines of force and the physical properties of the implant featured predominately. While the remainder of the lecture was inspiring and stimulating, as I had a dislike for physics, doubts began to enter my mind as to whether I would become an Orthopaedic Surgeon. Certainly the surgery would be rewarding but to have a chance of making any advances in the discipline this would be difficult without this branch of science.

Working at this intensity required periods of relaxation to recharge the batteries and remembering the lessons of second MB I went home occasionally to sleep and try my hand at golf. In September I took a weekend off to go to the wedding of an old school friend. Not having partied for some months I ended up somewhat the worse for ware in the front passenger seat of the bridal going away car. The bride and groom duly entered and commented that it was nice to get away from all the celebrations and be on their own at last whereupon I duly appeared and turned to face them. The car stopped and I was evicted but this incident having much amused them is still recalled today.

A short time later came the day of the exam. I was lucky in that after the grind of the previous seven months I was able to answer all the questions in the written papers. Attending for the orals I was nearly stymied in the pathology section when I was asked to name diseases carried by rats. I named one and then went blank. Luckily, with speed of thought, I stated that I knew there were more and if confronted with a patient bitten by a rat, I would make haste to the library and check the list, adding further that I never intended to work outside the country. The examiner smiled and I observed a tick put on his sheet opposite my name. Later that day we attended at the set hour in the Hall of the College for the names of those passing to be read out. I was one of the lucky ones joining the band of thirty percent passing at first attempt.

I retired to Alton for a few days relaxation but at the same time scouring the journals for a locum post and planning my next moves. To take the final fellowship I needed to complete a further year of training in recognized posts. I began by taking up a locum registrar post at the West Herts Hospital, Hemel Hempstead which having done no clinical work for seven months was a challenge as it was also a step up. However I needn't have worried as the two consultants for whom I worked were most supportive and once back into the swim of things my experience advanced, there being no disasters or untoward complications during my time there. There were no problems with adult patients but I found care of the young particularly challenging and vowed to attempt to gain experience in Paediatric Surgery during my training.

I applied for a Senior House officer post on the urological unit at St. Peters, Chertsey and once more I was successful being appointed to work on the unit with Mr. Mimpriss, whose main appointment was as a Consultant Urologist at St. Thomas'. He was an excellent teacher and was well supported by his registrar, Gordon Smart later a consultant urologist at Leicester. During the six months I became familiar with the instruments used in this discipline and gained experience in the more common procedures.

Just prior to starting at Chertsey, there had been a paper in the Medical Journals reporting the presence of a staphylococcal infection resistant to all antibiotics available at that time. It was labelled the Ward 7 Staphylococcus. When looking into this on arrival it appeared that patients had routinely been put on treatment without taking a specimen and sending it to the laboratory for culture and sensitivity. We changed the procedure so that no antibiotic was given until we knew what we were treating and what the sensitivities were. Within a short time span the Ward 7 bug was no longer a clinical entity. I adhered to this protocol as far as possible throughout my clinical career and believe that there wouldn't be the problems today with resistant organisms if more had followed this approach. One was certainly taught that this was the line to follow but then I can to some degree understand the pressure put upon doctors by patients demanding treatment. Oh for a bottle of coloured water or a placebo to overcome this attitude of mind. The body is a magnificent creation and certainly in ninety percent of the otherwise fit is capable of overcoming a viral or bacterial infection

One day, after a patient died I had to, in order to fulfil legal regulations, inspect the body prior to signing a cremation form. On arrival at the mortuary, the attendant said the undertakers had just left by the back gate but if I hurried I might catch the hearse near the front of the hospital. Somewhat out of breath I arrived just as it came into view. I flagged it down and having explained that I wasn't a body snatcher and what the problem was, the driver pulled into a nearby narrow lane with ten foot banks on either side. He opened the back and I performed the necessary examination. As the lid was closed I looked up and saw at the top of the bank a man as white as a sheet and eyes out on stalks who had obviously watched the proceedings. In view of his shock the situation was explained and on return to the hospital I called into the Secretary who sent a letter apologizing for the event. Nothing further was heard.

The mess at St. Peters comprised of staff from throughout the Empire or Commonwealth and had a most friendly atmosphere. On occasions there would be parties and on others, curry nights. Among my colleagues was a diminutive Maltese Gynaecological Registrar about five feet-three inches tall, a ball of fire and amusement and a jovial Indian Senior House officer a bulky man of over six feet doing orthopaedics. On curry nights they would go out into the hospital grounds the former with a rifle and the latter with a torch walking in step one behind the other, the Maltese leading. Rabbits were plentiful and when hit by torch light they tended to freeze and it would not be long before the team returned with the fare for the night which the Indian would then cook.

On one night off, I went to the Dorchester Hotel in London accompanied by a Nightingale now working at Chertsey and whose brother I had known at Thomas'. The evening was in aid of Queen Elizabeth's College for the Disabled of which my father was chairman. I remember the evening well for lady luck for the only time in my life was with me that night when I won a lucky ticket prize and five tombola prizes. I still wait for lady luck to return.

The six months passed quickly and it was time to move on. I plumped for an academic centre and after an interview by Prof. Ian Aird was appointed to the surgical department of Hammersmith Hospital (The Royal Postgraduate Medical School, now part of Imperial College), the location of which is possibly better known to some as it is next to Wormwood Scrubs prison. During the six months the Senior House officers, to widen their experience, moved at two monthly intervals to a different firm, each unit specializing in a different branch of surgery.

I started in September 1962 on Prof. Aird's firm together with a Chinese Australian, John Chong. The professor was a diminutive, highly energetic man with an amazing memory which had enabled him to write single handed a surgical bible entitled "Companion to Surgical Studies" first published in 1949. His contribution to surgery was immense having carried out the first kidney transplant in the UK where the donor and recipient were not identical twins. He led the team which developed the heart lung bypass machine. In 1953, he obtained public fame when he separated the Siamese twins, Boko and Tomo who came from West Africa. He was a man with enormous intellect. He used on occasions to enjoy going to a local pub invariably accompanied by senior surgical trainees. It was said that after a few drinks he could become contrary, belligerent and combative with other drinkers whereupon it was the duty of his surgical escorts to extract him from the pub and ensure him a safe journey home before mayhem arose. We had only one full teaching ward round with him before, being a depressive, he took his own life with an overdose at the age of fifty-seven. As Mr. Boggon at St. Thomas' had previously fallen by the wayside when I was his House Surgeon my peers began to rib me about being a surgical Jonah.

The professorial unit was then led pending the appointment of a new professor by a Scot, Tom Menzies a highly competent and knowledgeable surgeon awaiting a consultant appointment, which a short time later he obtained in Glasgow

Every Thursday morning there would be a surgical seminar presented by each surgical firm in turn with the highest down to the lowest in attendance. There would be more than forty attending and as the Hammersmith was a hive of academia and research most of the participants were spending a considerable time in the library when otherwise not occupied. These sessions could become an arena for one upmanship and scoring points over rivals. However the presentations and discussions were at the top end of the scale, an ideal step towards the final Fellowship, looming on the horizon. After two months I moved to the next firm of which Peter Martin one of top five vascular surgeons in the country and a founder member of the Vascular Surgical society was the leader. He was an inspiring man with large hands but a delicate touch who achieved excellent results with difficult procedures. I was influenced to the extent where I resolved to try later to seek a Registrar post with him at Chelmsford.

Late at night a group of us would meet in the Resident Surgical Registrar's flat and begin playing bridge at midnight. There were five or six of us including one who had been a county bridge player. Our abilities, with his instruction, subsequently advanced rapidly and we soon reached a level sufficient to give our mentor a reasonable game.

One evening a group of us went to the New Friends Chinese restaurant in the East End. John Chong, who spoke royal Chinese, was in the party and was designated to order the meal. The evening was memorable for two reasons. First John could not communicate with the Restaurateur as he only spoke royal Chinese while the staff spoke a different dialect. Secondly an army surgeon Major Moffat on a sabbatical from the Army for surgical training demonstrated that by eating slowly over two hours he was able to devour twice as much as anyone else present. It was a happy evening and a new experience.

After Peter Martin's firm, my final two months were with Richard Franklin also on at Kingston. He had trained at Thomas' between the wars and was an elegant neat surgeon with fine precise hand movements. He had performed in the nineteen thirties the first successful repair in a neonate of a tracheo-oesophageal fistula (malformed gullet with no connection to the stomach). He had a remarkable memory for conditions he had treated and was an excellent teacher, on many occasions recounting his experiences over a cup of coffee in the Sister's office at the end of a ward round. During my time on this firm a patient was admitted with a tumour of the small bowel. This was then discussed at one of the seminar sessions on a Thursday morning together with a complete resume of all known tumours of the small bowel. On another day I had a patient admitted for whom I was responsible. It turned out that this patient had been in contact with asbestos earlier in life and now had a mesothelioma in the left chest, a cancer resulting from this exposure. This case I had to present at the seminar, a somewhat nerve-wracking experience but at least I had good back-up from my more senior colleagues.

On leaving the Hammersmith I realized that my surgical knowledge had been taken to a considerably higher level which would auger well for the final exam. However there were gaps in other disciplines in my knowledge so I applied for and went on the final fellowship course back at St. Thomas'. This was a concentrated eight weeks, covering in addition to general surgery, other disciplines like neuro-surgery, orthopaedics and more minor specialities about which we needed to have a basic knowledge. We would have three or four lectures a day interspersed with clinical sessions when we would examine a patient and then have our presentational skills fine-tuned by the course lecturers. One of the highlights of this course was a weekend seminar at the Rowley

Bristow hospital Pyrford conducted by Alan Apley, a renowned teacher of orthopaedics and author of a textbook. He was a man whose organizational and brilliant teaching abilities far outweighed his skills as a surgeon.

After my customary weekend off before the exam, I entered the Hall in a good frame of mind. The papers turned out to be straightforward and easy, one of the questions asking about tumours of the bowel (thank you Hammersmith) and two others for which I was equally well prepared. The clinicals and orals all passed without mishap. At the table for my final viva, I happened to be wearing an old school tie, the examiner asked me as a final question why they still played soccer, rather than rugby, at Charterhouse. Having fallen over on the hard ground and broken an arm some years before, I was able to give a realistic answer from personal experience. He smiled and wished me well.

Once more at five o'clock we assembled in the Hall of the College and again I was fortunate to hear my name read out thus joining that relatively small number to have passed both parts first time, becoming a Fellow of the College in 1963, just four years after qualifying. The successful candidates then progressed to the committee room to meet the examiners and where once more I renewed my conversation about the old school.

On reflection afterwards, I thought the hardest exam I had ever taken was second MB. This was mainly because I had not exerted myself academically during the first fifteen months and had to learn everything for the first time (revision) in twelve weeks. This was followed by the primary Fellowship. The easiest was the final Fellowship for which I was best prepared soaking up knowledge in the subjects which most interested me.

I was now qualified to become a surgeon. I now had to go out and get the posts and practice to fulfil this ambition.

Chapter 5

Broadening my Experience

After the now customary short break, I searched the BMJ once more to find a post to tide me over until my next substantive job. My search was brief as the post that I had previously occupied at Hemel Hempstead was once more vacant.

Within ten days I was back at Hemel Hempstead. The West Herts hospital was the surgical side of a hospital divided and on two sites the Medical side being on another site two miles away. I was the senior member of the mess, my four juniors all being Indian over here to train and further their careers before returning to their native country. Three were Brahmins, the upper caste in the Indian system while the fourth was a Pariah, a lower caste. This led to friction and one of my tasks was to pour oil on troubled waters, to soothe the relationships.

At this time I purchased my first car, a red Triumph Vitesse with a white folding roof and numbered 50 HOT. This was a great joy and I was able to put it through its paces on the newly opened M1 Motorway.

Having worked at St. Peters, Chertsey and the Hammersmith, since last coming to Hemel Hempstead, on this occasion I was far more confident and had a much increased repertoire of operations about which I felt fully capable. However, I felt inadequate when it came to assessing the very young. Once more in the journal I found a post which appealed. I applied and was appointed to a junior registrar post at the Hospital for Sick Children at Great Ormond Street (GOS). While at the West Herts I performed a range of emergency procedures including operations for perforated duodenal ulcer and bowel obstruction.

One evening, we had a mess party and during the course of this a child of eight was admitted with abdominal pain. The Pariah Indian houseman was on duty and I sent him to clerk and examine the child. After ten minutes he reported back to me saying the kid had some pain but would last till the morning. I went to see the child myself and found a very sick boy with generalized peritonitis. I organized for him to go to theatre at the earliest possible moment and in the meantime went back to the party to talk to the houseman.

I saw red when having told him the true state of the child it became apparent that he did not have the same respect for life that I had been brought up to expect. On reflection later I realized that he had achieved much in life to get to his present position but he needed to adjust his values to conform to standards and work in the Western World. At the appointed time I went to theatre still inwardly seething, leaving the houseman behind. At operation I found a ruptured appendix with generalized peritonitis. Happily the child after a few days recovered and was fit to go home. To this day I shudder to think what would have happened if the lad had been left until the morning.

Before starting at Great Ormond Street I flew out to Spain to join my family for two weeks holiday.

I began at GOS in September on the Cardio Thoracic firm, the leader being David Waterston with his fellow consultant and assistant Eoin Aberdeen, an Australian. The Senior Registrar was Subra Subramanian, an Indian. David Waterston was a superb technical surgeon with a somewhat laid back attitude, while Eoin Aberdeen was a fund of knowledge and full of enthusiasm, but lacking the silky touch of David. Subra was also a fund of knowledge whose surgical skills were closer to those of David rather than Eoin. When I started in the post the repertoire of procedures for cardiac surgery was limited as this was just prior to the advent of open heart surgery. In addition to caring for the Cardio Thoracic patients, the junior registrars of whom there were four, also covered emergency admissions, being on call duty one week in four for these. I also had my own operating list initially assisted by the Resident Senior Registrar but soon being left to proceed unsupervised.

The nurses at GOS were similar in many ways to those at St. Thomas' but included a number from the British Commonwealth. One day when examining a male infant it suddenly peed on me. The

Australian nurse on the other side of the cot couldn't resist chortling with laughter. I turned the infant slightly and continued my examination. A few moments later it peed again but this time on to the nurse. I am afraid it became a case of he who laughs last, laughs loudest. After recovering from our bout of laughter which attracted others to the scene, the rest of the examination of the infant who had a blockage to outlet of the stomach (Hypertrophic Pyloric stenosis) giving rise to projectile vomiting was uneventful. Following a routine operation for this condition the infant made an uneventful recovery. GOS had staff flats for nurses near the hospital within bleeper range and John Kirkham, later a consultant surgeon at St. James Balham, and I would often be asked in for coffee following the day's work.

Towards the end of November there was a black tie dinner for medical staff. No one there will ever forget the occasion as the date was Friday 22nd November 1963, the day President Kennedy was assassinated. I learned this tragic news from Eoin Aberdeen during drinks before proceeding to the dining hall.

The doctors' mess at GOS was well endowed and some grateful person in the not too distant past had donated beer for the residents to have with their evening meal. As a resident I could invite a guest to attend. The food was excellent but invariably I would inform my guests often a brother that they had come on a good night.

Early in the New Year, the hospital acquired a heart bypass machine and after a few weeks we were ready to go. I arrived in theatre for these operations at 8 am. and assisted in preparing and setting the patient up for the procedure to be done . We began with a straight forward operation done expertly as always by David Waterston. The patient went on and off bypass without a hitch and then went to the Intensive Care unit (ITU) for recovery. This rapidly became a standard Wednesday happening, initially with one patient on the list, before with experience, doing two in the day. My duty following the operations was to remain in the ITU to monitor the patients until the following morning. If all was progressing well I would climb on a couch in the ITU around 2 am. and try and snatch a few minutes shut eye. I was lucky in that I could survive with catnaps. Although tired I was fortunate in that I could soon cope with routine problems but could always call Subra for anything untoward or that I had not encountered before. After doing my routine duties the following morning, I would retire to my room after lunch for my half day and sleep before jumping in my car, 50 HOT, to go out for the evening. The car number being so conspicuous not infrequently I would be told where I had been seen the previous evening.

The physical demands were heavy particularly when on for emergencies. Once more, emergency procedures were done late in the day and early evening. I would attend and assist and it wasn't long before I was doing emergency operations myself usually assisted by the Senior Registrar. I was pleased with this as I was performing at the Mecca of paediatric surgery in the UK and certainly one of the leading hospitals throughout the world. I became proficient at operations for strangulated hernias and for pyloric stenosis, a blockage to the stomach outlet causing projectile vomiting in neonates during the first four months of life. This was cured by doing a Ramstedts procedure which entailed dividing the muscle at the stomach outlet to allow drainage of stomach content into the small bowel and overcome the obstruction. I did operations which could be done at any hospital taking emergencies but assisted at those which should only be undertaken at a specialized centre such as GOS.

I much enjoyed my time at GOS and learned much, particularly valuing the experience I had gained in making diagnoses and performing routine paediatric surgery. Never again was I stretched in this field. Babies could come to you very ill but providing they were treated correctly their speed of recovery was amazing. During my time at GOS, one weekend when at home, I became ill with 'flu, which necessitated a week off. My resistance had probably been low due to the long hours worked. On occasions following open heart surgery this could be up to thirty hours at a time but after the week away I returned to the usual routine. On my departure from GOS after nine months, I learned that they were appointing two trainees to fill my post and some years later with the increase in open heart surgery this figure had risen to four!!!

Before starting my next post, I pondered whether I would like to become a paediatric surgeon. After some deliberation I decided against as consultant posts were limited and unless working at the Alder Hay hospital in Liverpool or at GOS, paediatric surgeons tended to work on their own rather than in Paediatric units with a number of surgeons. Inevitably there would not be the cross fertilization of interest and the associated stimulation of thought. While at GOS I found that when on site generally peoples' sense of humour was more at the paediatric rather than adult level and this I found unappealing for the long term.

I next moved to Chelmsford the attraction being Peter Martin, also on at the Hammersmith. The hospitals at Chelmsford at that time were on three sites: the Chelmsford and Essex in the City centre, St. Johns about a mile away to the south west and Broomfield in the country to the North of the City. The last at that time was evolving from being a Sanatorium catering mainly for patients with tuberculosis to also encompassing thoracic surgery for a multitude of conditions, including many patients having lung cancer. Today, Broomfield has developed on the abundant site into the main District General hospital the others being mainly confined to history.

The post was for eighteen months and rotated between three firms. I started with Bruce Pender, Thomas' trained, a general surgeon specializing in Urology and was based at the Chelmsford and Essex. Most of our operating lists were at St. Johns. The theatre at St. Johns was open plan and large with two operating tables so that two operations could be done in the theatre at the same time. On operating days, I would work at one table with Bruce working at the other. This made for easy communication and it was simple for Bruce to come across and oversee what I was doing or for me to slip over to witness some point of anatomical or pathological interest. A lot of operations usually fourteen to eighteen would be done in the day and would range from three to four majors each interspersed with intermediate and minor procedures.

One day, I was particularly pleased with the disciplines I had learned previously. A patient required a kidney removal and having placed the X-rays on the screen in the theatre, thus confirming the side to undergo operation, I went and scrubbed and gowned up while the patient was anaesthetized and put on the table. I arrived in the theatre with the patient prepared and towelled up, ready to go. I made a skin incision in the flank and got down to the muscle of the abdominal wall At this point I suddenly realized that the muscle fibres were orientated in the wrong direction and that I was operating on the wrong side. This was confirmed by a further look at the X-rays. Somewhat shocked, I closed the wound and had the patient turned so that the correct side was up. The pathological kidney was then removed without further ado. Next morning on my ward round I confessed to the patient what had happened. Much to my surprise he took it in his stride and was grateful that the error had been recognized before any damage had been done. Should the mistake not have been identified, this would have led to the patient's premature death and a large black mark ingrained on my brain for the rest of my surgical career. This near calamity reinforced on me the need to not only speak to patients before the operation but also to mark carefully the side and site of operation prior to the patient being put to sleep. This essential principal I adhered to for the next nigh on forty years of my surgical practice.

I was now doing an ever increasing variety of procedures and was learning to put my knowledge of anatomy into practice. I had always been impressed by the skills of the good technical surgeons with whom I had been fortunate to work. They had had the ability to economize in movement and get to a point in an operation in one skilled manoeuvre as opposed to four or even more by less skilled mortals. This had the effect of shortening the length of operation and hence the time under anaesthetic and so a speedier recovery. I was fortunate in that a lot of this had rubbed off on me and I remained throughout my career ever looking to see if there was a better approach to any manoeuvre. Throughout my time as a surgeon I learned technique, not only from revered seniors but from my peers and on occasions from my trainees. For the latter it was good for their moral and self esteem to know that teaching whatever their rank could be two way.

One day, just before the start of a list that had eighteen patients on it for the two tables I received a message that Bruce was ill. Rather than cancel any patients I decided to see if I could manage to do the operations single handed assisted by a houseman who was capable of closing wounds. The list consisted of six major procedures and twelve others. With two anaesthetists and two surgical teams and working virtually non stop, I moved from one operating table to the scrub room to the other operating table where gowned up I was ready to go. I managed to get through the list without a hitch just before seven pm. There was no putting my feet up as I was also on call that day for emergencies. After a rapid meal and assessing the intake I was back to theatre for four more operations, one of which was a major. Finally that day my duties were done. I had completed twenty two operations, seven of which were major without a hitch or mishap, all patients making uneventful recoveries. I was never to surpass, let alone equal that total again.

After six months I moved to Broomfield hospital. This rotation led me to witness the finest technical surgeon I ever watched and assisted. Geoffrey Flavell was on at The London Hospital and came to Broomfield hospital once a week to do a list. Also on at Broomfield was another surgeon, President at that time of the Thoracic Surgical Society now the Cardio Thoracic Surgical Society. This surgeon was competent. He would do a pneumonectomy (removal of lung) in two to two and half hours whereas Flavell nicknamed Bugsy would do the same procedure in thirty to forty minutes. I watched and assisted both with much interest and soon realized that Bugsy was doing in one movement what the other was doing in many. His knowledge of anatomy in the chest was comprehensive and his movements deft. Watching him was watching poetry in motion and I learned much. Again recovery after operation was related to time under anaesthetic.

Part of my job at Broomfield was to assess patients before operation. At that time I was smoking a mixture of cigars and cigarettes. Many of the patients coming for surgery had lung cancer and all except one was a cigarette smoker. The unfortunate one was a London policeman who had spent much time in central London on point duty directing traffic and at the same time inhaling fumes. At that time, inhalation of toxic fumes was not recognized as an occupational hazard.

The pace at Broomfield was more leisurely and I for a while contemplated doing further study to become a member of the Royal College of Physicians. However this did not progress far as I was away from main stream medicine and as time passed the idea became impractical. The doctors' mess at Broomfield was pleasant and one of the trainees was from Bombay where he had been the University Badminton champion. There was a large hall at the hospital for community activities which contained a court. Playing two to three times a week I was able after four months to give him a game although at first I had been given a real run around.

At Christmas, having accomplished ward duties including carving a turkey I drove home early in the afternoon. In 1965 there was not a car to speak of on the road and I arrived at my destination and joined the family late in the afternoon. As previously we were joined by Miss Walker the matron at Treloar's and my Father's then Registrar and wife. The break rapidly passed and I was soon back at Chelmsford. In view of my findings concerning patients with lung cancer I determined to ditch smoking of cigarettes. On 31st. December I smoked 30 and in addition some cigars. I could not stand the thought of another smoke the next day and since that time not another cigarette has passed my lips.

Having spent six months at Broomfield I returned to the Chelmsford and Essex, confident now that I could cope with any thoracic injury and having widened my experience for general surgery. I was now working for Peter Martin and an orthopaedic surgeon, as trauma, such as broken legs and fractured hips, was admitted to the emergency ward. Peter Martin was great to work with. He was one of the top five vascular surgeons in the country skilled with very large hands but surprisingly dextrous. He was also a thinking surgeon and would emphasize differing approaches to ensure that a patient had a sufficient blood supply to ensure limb salvage rather than succumbing to amputation.

His results were excellent and he was a model as to how to cope with a life or death situation such as a ruptured aneurysm (a localized blow out of an artery) of the abdominal aorta (the main blood vessel coming from the chest to supply the abdominal contents and lower limbs). Seventy percent of patients with this condition never survived long enough to reach hospital while for those that did the survival rate in the best hands was between fifty-five and seventy percent whereas in others it could be as low as twenty to forty percent Once the patient was asleep all would be ready before the anaesthetist relaxed the patient. As the relaxant worked the abdominal wall would go limp and no longer would pressure be applied to help control the bleeding vessel and blood loss. At this stage the surgeon had to work fast to enter the abdomen, assess the situation, dissect quickly to get round the aorta above the aneurysm in the upper abdomen and apply clamps to stop the leakage of blood, potentially torrential. Once this stage was reached all was under control and if needs be one could retire for a cup of coffee before returning to fashion the reconstruction. Much rubbed off on me as a result of assisting Peter at these operations and I found when confronted with this condition later, my adrenaline would squirt and I would become ice cool and focused working rapidly until the clamps were applied and I could relax with the situation under control. During the six months I performed a lot of vascular procedures and learned techniques that sustained me for the remainder of my career.

On the orthopaedic side I became proficient in pinning fractured necks of femur, a condition seen commonly in the elderly and particularly women suffering from osteoporosis. In view of the ages of the patients, the least time spent on the operating table the better. I put into practice a technique I had learned from one of my Father's registrars, Carlo Biaggi, some years before. He had taught me how to reduce the fracture and with an Xray confirming a satisfactory position how to insert a guide wire at the correct angle. Having inserted the guide wire I scrubbed and gowned and then through a very small incision inserted the pin (Smith Peterson) over the guide and completed the operative procedure to stabilize the fracture in less than fifteen minutes. The other common condition requiring treatment was a fracture of wrist (usually a Colles) which under outpatient anaesthesia could be reduced, set and put in plaster.

One day a patient was admitted with breathing difficulties. I examined and assessed him and judging that there was no immediate problem went to dinner. During the course of the meal I received an emergency bleep and rushed back to the ward to find the patient blue and unable to breathe. There was no time to take him to theatre or scrub, so I called for a scalpel cut into his neck made a hole in the wind pipe and inserted a tube to allow breath, before taking him to theatre to tidy up. This patient went on to make a good recovery. Two days later the microbiologist rang me about an unusual organism he had grown and wondered if I could throw any light on it. I remembered I had been eating blue cheese before I rushed to the ward and that there had not been time to wash my hands before my attempt to resuscitate. The organism proved to come from this but happily gave no problem.

While at Chelmsford, I went to Italy for a holiday, staying on the Amalfi coast. On the Saturday following my return I suddenly developed a rigor which did not settle with normal medication and necessitated my admission to the hospital. The rigor continued but it was not until the next afternoon when the hospital matron was visiting that she noticed that I was jaundiced. I remained in hospital for two weeks before going home to convalesce from infective hepatitis. On the first Saturday home, I went with my father to watch racing at Newbury but returned to the car somewhat exhausted after the second race. Two weeks later we returned to the same racecourse and by now took the afternoon in my stride. On returning to Chelmsford, I was relieved of night duties for two weeks but was otherwise back to full activities. Because of the illness I had been banned from having any alcohol for six months. I found that when attending parties I was soon carried along in the party spirit and that as a party progressed, peoples' tongues would become far looser and that not having had a drink I could remember all that was said the next day. I gleaned information that normally would not have registered and learned another lesson for life.

Peter Martin in addition to being an excellent surgeon and teacher was also a good host. To recharge his batteries, he kept a boat at Burnham on Crouch and on occasions sailed across to Holland to attend meetings. Once a year, he would take his surgical team and theatre staff out for a day's sailing, in my year a most enjoyable occasion. After a day's operating we would often retire to a local pub to relax and talk over the day's events. On one occasion whilst on the way there, a stranger came up and said Mr. Martin I must shake your hand and thank you for what you did for me. Peter spoke to him for some five minutes before moving on. When we asked him who the patient was, and what he did for him, he said he hadn't a clue but during the five minutes he had been able to give the impression that he knew by subtle questioning all that happened and what he had done for him. The patient had moved on very happy and again I had learned another lesson for life.

After some twenty months at Chelmsford and having gained much in knowledge and experience it was time once more to take the next step up the ladder. I left Chelmsford highly satisfied with my time there and requests by consultant staff to apply for any job which might come up as I would be most welcome to return. But for that part of Essex being very flat and to my mind the terrain dull and uninteresting I would have put in for a consultant post there some years later.

Competition for Consultant posts at this time was fierce so to have the opportunity to be appointed to a good and desirable post it appeared necessary to add more to one's CV to be able to stand a chance of getting on a short list let alone getting appointed. By this stage I had worked in leading hospitals in and around London and had been in a position to hone my surgical expertise but as yet had no research or academic experience under my belt. I had not written a paper and the cry "publish or perish" was reaching my ears.

A post of Surgical Registrar back at my alma mater, St. Thomas' was advertised so I applied and was delighted to get appointed.

Chapter 6

Back to the Alma Mater

I was only the second trainee surgeon to leave and return to St. Thomas', all previous incumbents before becoming consultants, having remained at St. Thomas' throughout their training. I was indebted to Donald Reid, later a consultant surgeon in Brighton, who some years before was the first, for inspiring me, while I was the Senior Surgical Casualty Officer, to make this move. Without this experience I would not have seen and worked with some of the finest thinking and technical surgeons around. I found it an advantage to be in a position to quietly bring new techniques to St. Thomas' to those honed by generations of inbreeding while at the same time being in the position to very much take onboard what they had to offer.

Although returning with considerable experience it was nevertheless back to the bottom of the ladder. By chance, I began back on the firm where I had worked as a houseman some years previously, with Bob Nevin and Lyn Lockart-Mummery, the latter having been the first outsider to be appointed to the surgical staff at St. Thomas'. He was also on at St. Marks, the Colo-Rectal Hospital now incorporated and part of Northwick Park Hospital. He had brought a new dimension being sound clinically and a most delicate and elegant surgeon and his work was a joy to behold. The Senior Registrar was Brendan Devlin originating from Dublin, knowledgeable, outspoken and not afraid to express his opinion often to his detriment. He was an adequate technical surgeon but a good mentor impressing on me the need to get into print. I learned a lot particularly from Lyn relating to surgery to the lower bowel which was again to stand me in good stead for the years ahead.

At this time, my brother, John, five years my junior was a house physician on one of the medical firms. One day, when on duty for emergencies I received a call from him saying he had a patient he would like me to review. This patient was a sixty six year old female who was admitted under the Physicians in cardiac failure resulting from the ravages of Rheumatic heart disease years before. In addition to valve problems, she had an irregular heart beat (auricular fibrillation) which can predispose to blood clots forming within the heart. Some forty eight hours after admission to hospital, she was referred while on treatment for her cardiac failure because she had developed right sided abdominal pain. I made a diagnosis of acute cholecystitis or acute appendicitis should the appendix happen to be in an abnormal anatomical position. In view of the patient's general condition and the heart failure it was decided to treat her conservatively and observe. Three days later she developed generalized peritonitis and it became imperative to operate. As the patient was being put to sleep the anaesthetist recorded that a pulse in the right wrist had disappeared. On opening the abdomen, bile was seen to have leaked through a somewhat necrotic wall of the gall bladder. Usually, when there is inflammation of the gall bladder, as with acute cholecystitis, surrounding tissues become adherent forming an inflammatory mass to wall off the infection. Both the gall bladder and the surrounding tissues require a blood supply for this to happen. At this operation the appearances were unusual in that there were no surrounding adhesions. The gall bladder was removed and peritoneal cavity toilet done. After operation it was noted that in addition to a wrist pulse having disappeared the left groin pulse supplying arterial blood to the left leg had gone.

The clinical state had now become evident. It appeared that the patient was sending off showers of clots (emboli) from her heart which were then occluding peripheral arteries. The left leg became pre-gangrenous so I took the patient back to theatre and under local anaesthesia extracted the clot restoring blood flow to the leg. The patient in addition to her previous medical treatment was placed on ante coagulants and went on to make a full recovery.

I next turned my attention to the histo-pathologist and discussed the clinical situation with him. He was able to confirm that indeed the gall bladder was necrotic due to its artery of supply being blocked. Search of the medical literature showed this pathology to be rare if not unique. It was only a short time before I had my first publication with my brother, being a co-author having referred the patient. Some weeks later I received a letter from a surgeon in the States saying that he had always thought that this might be a clinical entity and had been on the lookout for a case. He duly confirmed that this was the first recorded case in the world literature.

For one operating list we had a guest surgeon, Norman Tanner, the doyen of gastric surgeons, who was on at St. James' hospital, Balham and the Charing Cross. St. James became a Mecca for surgeons of international repute to visit, so it was a privilege to welcome him to our theatre at St. Thomas' to demonstrate his skills. With the inbreeding that had been prevalent at St. Thomas' there was much to admire and learn.

Brendan Devlin was eager to determine if patients undergoing gastric surgery and who sustained inadvertent damage to the spleen requiring removal of that organ, had suffered any long term detrimental effects. This study involved going through St. Thomas' Hospital records for ten years, pulling all the notes of those patients who had had a splenectomy. During this time two hundred and seventy five splenectomies were done of which twenty five percent resulted from splenic injury mainly associated with surgery to the adjacent organ, the stomach. Patients suffering inadvertent splenectomy had a noticeably prolonged post operative stay due mainly to minor complications, but fortunately did not suffer any long term sequelae. My contribution to this paper was to retrieve the records of the two hundred and seventy five patients and extract the information required. For me this was a good lesson in the discipline required while doing research projects.

The surgical registrars were rotated once a year and after four months I moved to the Royal Waterloo hospital down the road from St Thomas'. There I ran the surgical side working for the first time with Frank Cockett and the Reader of the St. Thomas' surgical unit, both of whom did weekly lists in this hospital. Frank Cockett was one of the five founder members of the Vascular Surgical Society another member being Peter Martin with whom I had previously worked. The Reader was Norman Browse who came to St Thomas' on appointment and remained there throughout his surgical career subsequently becoming Professor of Surgery and ultimately President of the Royal College of Surgeons.

Frank Cockett was acknowledged as one of the leading varicose vein surgeons in the country if not the world. During his training he had studied the venous drainage of the legs. He had found that in patients with incipient leg ulcers, there were damaged valves in the calves. Instead of ensuring venous flow only from the superficial veins to the deep veins, the valves had become incompetent. This led to venous stagnation in the lower leg, with increased pressure and poor oxygenation of the subcutaneous tissues, resulting in necrosis and ulceration. Pressure bandaging would overcome this deficiency and enable healing. With the ulcer healed, the patient could undergo venous surgery and severance of incompetent perforator veins to prevent a recurrence. The operation subsequently became widely known as a Cockett procedure.

After a short time in post, having had his operation demonstrated and with my previous experience, I was given a free hand to attack the daunting waiting list. During the course of my year there I operated on some five hundred patients, managing to make a large hole in the waiting list. On Thursday mornings Frank would instruct me to start the list, which usually contained at least one major arterial case. The operations demanded that the anatomy was first displayed, before an ante coagulant was given, whereupon the artery could be opened and the necessary procedure done. At first Frank would arrive almost at the moment when the ante coagulant was given but as time went on I got to do more, until there were times when I would have completed the first operation before he entered.

Norman Browse came from the Westminster hospital. He was a talented surgeon and one could see he was going to go far, which indeed he did. During this time I lived in the Hospital and was available for emergencies. I could get out and now that exams were behind me I no longer had to be single minded about my studies. Being destined to almost certainly become a Consultant I branched out and began to have a social life. During the next year or two I met and much enjoyed the company of a number of Nightingale nurses. I almost became engaged to one but she decided to take time out and go with a friend to Australia. The time she was away extended and I was left with uncertainty with regard to the future. However friendships continued to flourish.

One day, I was asked to see a patient on the psychiatric unit within the hospital. The patient was a female aged twenty suffering from Anorexia Nervosa. On admission to hospital her weight had been 39 Kgs whereas fifteen months earlier she had weighed 54 Kgs. She was put on medical treatment and her appetite immediately returned to normal. Five days after starting treatment she developed severe abdominal pain. On examination she was shocked and her abdomen was grossly distended. She was taken to theatre and on opening the abdomen with the drop in pressure, the stomach spontaneously ruptured, soiling the abdominal cavity. The anterior surface of the stomach was necrotic and gangrenous. There was sufficient normal stomach tissue at the proximal and distal end to be able to remove the pathological area and reconstitute the stomach joining viable stomach to viable stomach. The patient went on to make an uneventful recovery going home nineteen days after operation.

This patient, having eaten virtually nothing for fifteen months, began within a day of starting on medical treatment to eat normally. The muscle of the stomach wall had wasted due to inactivity and could no longer contract in a normal manner. This led to gross dilatation of the stomach which precipitated clotting in the vessels of the stomach wall leading to gangrene. This condition had previously been reported and had been seen when starving inmates of German prisons were released towards the end of WW2. Four ex prisoners are recorded as having suffered from this condition as a result of eating normally on release before it became apparent to the liberators that eating initially required supervision and control. This patient had eaten voraciously after commencing treatment and had suffered the same pathology. This case history was again published and served as warning for the rapid treatment of patients with anorexia.

The year at the Royal Waterloo passed rapidly and once more it was time to move on. I now had worked with two of the leading vascular surgeons in the country together with an up and coming one. By that time, I had had the opportunity to perform a large number of operative procedures and gain experience far in excess of my peers.

The results achieved supported the faith they placed in me by allowing me to proceed.

In September 1966, I moved to St. Helier Hospital Carshalton, on the St Thomas' training circuit where I worked with Aubrey York Mason and John Edwards, previously a senior registrar on the Surgical Unit at St. Thomas'. This meant a daily commute from where I was staying with John Fergus a fellow contemporary training to be an urologist. After Norman Tanner, Aubrey was recognized as the next best upper gastrointestinal surgeon in London. There was much to learn from him about upper gastrointestinal surgery when, prior to the development of medical treatment some ten years later, surgery for duodenal ulcer and its complications, was common.

Aubrey operated on a number of colleagues and members of the medical profession during my time with him. This could be nerve wracking, operating on someone you might know well. He instilled in me the importance of doing your standard operation on these patients as the reason they came to you was because they were aware of your results as well as your abilities. All went well until one day he was operating on an anaesthetic colleague with cancer of the head of the pancreas. He performed a Whipples procedure, removing the head of pancreas and completing some demanding anastomoses to re-establish gut continuity and bile drainage from the liver. All appeared to have gone well technically but some hours later complications arose. Despite re-operation the patient's condition deteriorated and he died. Aubrey was obviously much affected and upset by this. It was at this moment in time that I realized fully for the first time the stress that surgeons could be put under following their daily practice. Your mind is charged when operating on a friend or colleague with all the things that might go wrong and inevitably you begin to operate under a degree of tension. I was lucky in that once the skin incision had been made, these thoughts were put to the back of my mind, as I became engrossed and concentrated on what I was doing. The operation then continued normally without the preoperative apprehensions. The patients benefited and I after the initial pangs and worries would operate as if it was just another case of such and such.

John Edwards, a Welshman, was a general and vascular surgeon so on this side of the firm it was mainly covering ground that I had already done at St. Thomas'. Aubrey and John were both excellent teachers and led good weekly surgical firm meetings and discussions. Also attending these were a number of other trainees from a nearby hospital. One of these that I got to know was David Rosin who was appointed later to the staff of St. Mary's Hospital Paddington. Our paths were to cross again some years later.

I moved next to the Lambeth hospital which had once been a workhouse prior to being taken over by the LCC. Following the creation of the NHS it had been incorporated into the St. Thomas' set up with trainee medical and surgical staff rotating there and students attending for teaching. The consultants were inherited. Gordon Ungley was also on at the Westminster hospital while Mr. Eason was a journey man ex LCC surgeon. Pasty Barrett, an Old Etonian, a brilliant speaker and eminent Thoracic surgeon who had been at St.Thomas' throughout his surgical career, attended to give teaching rounds.

Gordon Ungley was a neat, dapper and precise surgeon who ensured that one was accurate in the use of terminology. With Mr. Eason I had one of the more embarrassing moments of my surgical training. We were operating on a patient with bilateral varicose veins, both legs being similarly affected. After fifteen minutes, with the experience of the previous year, I had completed my leg and was instructed to go and get my lunch. Some thirty-five to forty minutes later I returned to find the patient still on the table with the consultant still operating, the operation on that leg still incomplete.

In June 1967, I was contacted by the sister of City ward in the main hospital. She had previously been ward sister of Nuffield ward, where I had done my early jobs and whom I had known for six or more years. She asked me to complete and run for her team in the ward relay race at the annual St. Thomas' hospital Sports Day at Cobham. On arriving there I found that one of the team members was a friend of my brother's girl friend at that time. I was lead off man for our team and by dint of anticipating the starter's gun I arrived at the first change over in the lead to pass the baton to the first female in the team, a nurse from City ward who had been sent down to make up the team. This nurse was to become my wife. That evening we went to the Sports Day hop in St. Thomas' House. Mary for the first and only time in her life asked for a whisky and ginger ale to drink. Despite our age difference of twelve years, our relationship flourished and it was not long before I asked her home to meet the family. What might have been daunting for some she took in her stride as she was one of six, herself and five brothers. Sadly one had died while away at school in Dublin where he had developed hepatitis. Completely understandably this had left a scar in her family particularly with her mother.

I next went to Lancashire and met Mary's parents. Her father was a GP, who had trained at the College of Surgeons in Dublin and her mother had been a nurse, a lovely softly-spoken person, kind, thoughtful, considerate and loving. She had done a lot for the local community having been mayor. She had managed to get many stone houses in the valley refurbished, a far better solution to the housing problem than pulling them down and putting up jerry-built modern abodes. Her father was an excellent GP, who didn't stand for nonsense or possible skivers. He would prescribe antibiotics if indicated but realizing most people had a self-limiting illness such as 'flu or a cough doing the rounds of the locals, he would give them a bottle of coloured water to take as prescribed, telling them at the same time to stay out of the cold and that they would be better within forty-eight hours. He recounted how patients would come back the following year with similar symptoms asking for a further bottle of the green medicine. With my previous experience at St. Peter's hospital, Chertsey, of the ward seven staphylococcus, I fully understood and supported this approach. If more trained staff had adopted this attitude, there would not be the problems today that there are with antibiotics when organisms are becoming resistant to treatment. By pandering to the public and giving inappropriate medication for the treatment of the mind, doctors are running out of therapeutic options. Yet another lesson for life was reinforced. My future father-in-law was a good raconteur and had a great memory for past clinical experiences, particularly when in Ireland, with which he would regale us. Often there would be a lesson in his utterings for long-term people management.

My father-in-law to be had a passion for good wine, about which he was knowledgeable and helped to educate my taste as well as increasing my experience. He would go a long way for a bargain and once when in Scotland drove halfway across the country to obtain some Calon Segur. On arriving at the shop off the beaten track the lady behind the counter discounted the product already at a low price because she thought it old. Little did she realize that it was "à point" and just reaching its best. The car came back to Lancashire well loaded!

By now I had changed my car to an MG. One evening I was running Mary back to her nursing home on Cheney Walk, Chelsea, having left instructions for the theatre staff to prepare a patient for surgery and to get him on the table for me to do an appendicectomy. I was slightly delayed and was speeding back to the Lambeth when I was gonged by an unmarked police car. I explained the situation to the officer and that a patient was on the operating table anaesthetised waiting for my attendance. He let me go and I drove off at a similar speed followed by the policeman coming to the Lambeth hospital to verify my comments. Having done so, I heard nothing further. I suppose I learned that if you are going to speed you need a good reason.

Towards the end of 1967 I got engaged to Mary. She came from a devout Catholic family whereas I had been brought up in the other arm of Christianity, the Church of England. This difference was discussed at some length with the result that I decided to change to the Catholic faith so that for any children in the future there would be unanimity of belief. It was necessary to receive instruction and I was fortunate that my future father-in-law found a priest based at the Catholic Radio and TV broadcasting centre at Hatch End in North London. Father Michael Childs, the deputy to Father Agnellus Andrew, had been at Cambridge as an undergraduate and had been married. His wife and baby had died in childbirth and following this sad event he took Holy Orders. He was highly intelligent and a good communicator, preacher and teacher. He didn't attempt to spoon-feed me with doctrine. We would meet for dinner at Hatch End or in central London and then discuss differing aspects of the Faith and the differences between the Church of England and Catholicism and the significance of these variations. Father Michael was an inspiring man with a good sense of humour and I much valued his counselling.

A date was fixed for the wedding in May 1968. In the intervening time, we found a first floor flat in Albert Bridge Road, opposite Battersea Park. This was well situated for travel to the hospital and within walking distance of the Kings Road, Chelsea very much a buzz place at that time.

We married in Lancashire in May on a very cold but dry day, the service being led by Father Michael who with his background and a very mixed congregation gave an inspiring address, well received by everyone. The ceremony was followed by a sumptuous reception in a marquee, I was pleased to note well heated to overcome the unfriendly conditions. This was a great day for Mary's parents, no expense being spared for the marriage of their only daughter.

After honeymooning in the Canary Islands, we returned to London and our work. Mary was the first married trainee nurse to be allowed to complete her course at St.Thomas' to become a State Registered Nurse (SRN) and obtain the cherished and much respected Nightingale badge. All previous trainees getting married had been asked to leave. I often wondered subsequently because I was surgeon to the Nurses Clinic whether this influenced the rewriting of this paragraph in the Nightingale School rule book!!

By this time, I had reached the top of the pile of trainee registrars. To ensure that I was fully qualified to get a good consultant post I needed to do some research with a view to getting a Master of Surgery (MS) degree. A post of research Registrar became available, which was to be supervised and mentored by Frank Cockett. This was a new position. Previously research had been done on the surgical unit under the supervision of Professor Kinmonth who was a world authority on the lymphatic drainage of the lower limb. He was running a long-term laboratory project studying the lymphatic drainage of the lower limb, each incumbent slowly advancing the knowledge and understanding of the mechanism of this system. Frank Cockett and Norman Browse had already identified that relationships between the professor and me would almost certainly not be harmonious. He was a ponderous, pedantic surgeon who would have a choice of ten instruments to do a particular task whereas one would do. He was the antithesis of the way I had been trained and would operate and, therefore, almost certainly our relationship would have been a disaster waiting to happen.

I was appointed to become Frank Cockett's first research Registrar.

Chapter 7

Spot the Clot

And so to research. There was room for me in Frank's office which also housed the firm secretary. This was situated on the first floor of the main hospital towards the west end. I met up with Frank to discuss my project. At first this appeared straightforward but soon I realized that it required much trawling of the literature and research if I was going to get anywhere.

Frank just said to me "Too many people are dying from Pulmonary Embolism, see what you can do about it". Five in every hundred hospital deaths were due to this catastrophe and occurred in patients many of whom were expected to return home to a normal way of life.

A typical scenario would be as follows. A patient may have had a major and possibly lengthy operation and would appear to be well on the way to recovery. He might have had a minor blip in temperature or a little ankle swelling or even some minor discomfort in the calf. These somewhat nebulous signs were rarely found. Not infrequently, some seven to ten days after operation, he would get a call to go the toilet and while there and possibly straining would suddenly collapse and die. A post mortem would be performed and the pathologist would examine first the site of the recent operation and find all well and healing soundly. He would next examine the heart and lungs and find on opening the pulmonary artery that it was occluded by an often lengthy and large coiled clot. This had travelled as an embolus after breaking off from its source up the venous system from large veins in the legs or pelvis to totally occlude the circulation beyond the right side of the heart. Most patients died almost immediately following such an episode, with only the small minority surviving long enough to undergo investigation and treatment. The pathologist would next examine the large veins in the pelvis and the main veins in the legs down to the upper calf. A typical finding would be the presence of clot extending from calf veins into the vein behind the knee where it would be obvious from where the clot had broken off and travelled as an embolus up to the heart.

It would have passed through the right side of the heart and out into the pulmonary artery where it had impacted, totally obstructing blood flow to the lungs and causing the cessation of circulation.

From these observations I needed to confirm the weight and volume of clot required to occlude the pulmonary artery and cause near death or death. I started to trawl the relevant literature and it was soon obvious that many eminent clinicians had been researching this problem and writing learned papers for more than one hundred years. How was I to get anywhere following such a galaxy of talent unless I had tools and instruments not available to these clinicians? At this stage, I concentrated on determining the prevalence of the condition and with advances in micro technology looking to see whether an instrument or tool could be developed to fragment or remove the embolus in those lucky enough not to be killed instantly.

On researching the literature, I found that an embolus had first been described in 1686, but the Italian author, as with subsequent authors, failed to identify from where the clot arose and how it reached its point of impaction. It was not until the mid 1800s that Rudolf Virchow, an eminent German scientist and pathologist, first described the mechanism of thrombo-embolism linking the formation of peripheral venous clots with their propagation and then breaking off, and travelling in the venous system to the heart to occlude the pulmonary artery. From then onwards, there were sporadic papers reporting this condition.

Following the turn of the century much was written. It was variously recorded that, at post mortems in learned centres, five per cent of all deaths were due to this condition. Michael De Bakey, the eminent American vascular surgeon, in a mammoth review of the subject in 1954 reported that fifty percent of patients suffering from fatal and non fatal emboli did not have any clinical signs prior to the episode. If suspected, it was possible to inject dye into the peripheral veins of the legs and view X-rays to see if clot was present. Because it took time, X-ray investigation was only suitable to confirm or deny the presence of clot. It could not be used as a screening modality. During the 1960s, despite increased awareness and increased prophylactic treatment for patients, there appeared to be an increase in the number of patients dying from this condition. The Oxford Regional Health Authority, one of the smallest Health Authorities in England, now no longer in existence, after reviewing its post mortem findings in 1968, calculated that 148 unnecessary deaths occurred in one year; that is 148 patients who had been expected to get back to a normal life on leaving hospital.

Although attempts had previously been made, it was 1924 before a patient, suffering a pulmonary embolus, survived long enough to be taken to theatre and to be operated on successfully. There, his chest had been opened and through a small incision in the pulmonary artery the embolus extracted and the wounds closed. Between 1924 and 1958, only thirteen more successful cases were reported in the world literature. Following the development of cardio-pulmonary bypass surgery many more patients survived, but it was appreciated that only twenty percent of patients suffering a massive pulmonary embolus survived long enough to undergo surgery. In these, the survival rate with surgery was only about fifty percent.

I next retrieved and analyzed fatal emboli collected from the mortuary at St. Thomas' and the local coroner's mortuary at Southwark. In all I collected twenty-five specimens and these on measurement of weight, length and width confirmed that the emboli had originated from the lower thigh or the larger veins nearer to the heart.

Realizing that any progress was likely to arise from new inventions I arranged to go to the Atomic Weapons Research Establishment (AWRE) at Aldermaston knowing that it was to the forefront of new technology. I discussed the possibility of inserting a macerator and threading it up the venous system to fragment the embolus in the Pulmonary artery. It was soon apparent that this was a no-goer. I next looked at using high energy ultrasonic waves to disrupt, diminish and fracture the embolus. While this had an effect in vitro, when energy was applied directly to the embolus, it was ineffective in studies when the energy had to penetrate an arterial wall before hitting the embolus.

Having drawn a blank with my early research, I now sat down and had a rethink. It was not long before I realized that I was putting the cart before the horse. I had been looking to treat a patient after an event in which some eighty percent of patients were dead before any treatment could be instituted. It was now that I understood that I had to look at the source of an embolus and hopefully diagnose thrombosis which could be treated before a fatal episode occurred.

In the few cases where deep vein thrombosis (clotting) could be diagnosed, and confirmed by improved radiological techniques and with the availability of anti-clotting agents such as heparin, the outlook improved. In other patients, not suitable for anticoagulant treatment, a vein could be tied off or plicated to trap any embolus and prevent onward progress to the heart. Despite these measures, patients continued to die at the same rate emphasizing the lack of progress in making a diagnosis and the need to develop a simple screening test to overcome this silent killer.

A colleague, David Negus, had developed an isotope screening test whereby a radioisotope attached to fibrinogen, a constituent of blood, could be injected into a patient's vein and hence into the circulation. The calves would then be scanned on a regular daily basis. If a clot was forming in the calf the fibrinogen would be incorporated giving a hot spot on scanning. The presence of clot was then confirmed by radiology and treatment given as appropriate. This form of investigation was ideal as a research tool but suffered as it required a technician to perform and was a somewhat lengthy procedure, taking twenty minutes for each examination. It was not therefore suitable for screening large patient populations.

Analysis of patients dying from pulmonary embolism in four years at St. Thomas' identified that there had been one hundred and eleven deaths, giving an overall mortality of five percent from embolism. Of these sixty percent could have returned, but for the fatal incident, to a normal way of life.

About this time, some six months after commencing my research, I read a paper from America stating that the authors had used a low energy Doppler ultrasound scanner to evaluate venous flow in the legs. They used a 5MHz machine which enabled venous flow in the leg to be examined from the calf to the groin. The Doppler effect is best explained by a car or a train coming towards you with its horn blowing. As it comes towards you the sound is high-pitched, as the sound waves are compressed. When it is past you the sound waves are lengthened giving a drop in pitch. This effect is caused by the movement of the sound source. Using this technique, movement of the source being scanned gives sound, while with no movement, there is silence. I suddenly realized that this might be the way ahead, the technology not previously having been available.

I obtained two ultrasonic machines, one with a 5MHz and the other with a 2 MHz output, and it was soon obvious that the performance of the Sonicaid D205 was superior and it was therefore adopted. This apparatus consisted of a portable box weighing seven pounds and contained a transmitter operating at 2MHz, a receiver, an audio amplifier and a loudspeaker. The box was connected by a lead to a transducer which contained two ceramic crystals, one of which when excited emitted ultrasound while the other acted as receiver for the reflected back scattered sound. A recorder could be plugged in to the back of the box to achieve a permanent write out of each examination.

I remembered that Sonicaid machines were used in the maternity department of the hospital to monitor the foetal heart rate. I managed to borrow one for a few hours and took it to our office when I set it up and turned it on. After getting on the couch, I applied olive oil, as a coupling agent, to my groin and placed the transducer upon it. I immediately heard the roar of blood flow from the underlying artery and then sliding it less than one centimetre medially heard a lower intermittent sound. I discovered if venous flow in the leg was accelerated the intermittent sound became a loud whoosh. I extended my self-examination placing the transducer above the groin crease and then over the lower abdomen. In all positions I was able to get an increased sound on augmentation of venous flow.

It was soon found that if a squeeze was applied to the calf a roar could be heard from the loudspeaker. This was the Doppler effect in action. The squeeze on the calf gave a rapid acceleration in the rate of venous flow of the blood from the calf, shortening the length of the sound waves which was picked up by the receiving crystal. If the vein was blocked or occluded there would be no acceleration through the vein and silence in the loudspeaker.

A shortcoming in previous authors' works was that they could only examine flow from the calf to the groin. By changing the frequency of the scanner from 5MHz to 2MHz and obtaining greater penetration, it proved possible to examine venous flow from the calf to the upper abdomen thus ensuring all possible sources of life-threatening embolism could be investigated.

With these preliminary findings I contacted Sonicaid in Bognor Regis, who immediately asked me down to talk through this possible new application of their device. Following further discussions with Frank, I visited them and talked the project through with members of their staff.

They immediately saw the possibilities and potential of what was discussed and offered to back the research and supply me with any necessary equipment to fulfil this aim.

To ensure repeatable and reproducible results by others as well as myself, it was necessary to write a protocol for patient examination. Patients were examined in bed sitting at forty-five degrees with the ankles supported on a pillow to allow the calf veins to fill. To get a standard calf squeeze I devised a cuff compressor to be applied round the upper calf. By releasing a clip, the calf could be compressed to a standard pressure thus ensuring uniformity in examination. Once practiced with this procedure, each patient examination was taking about five minutes, a time which would enable screening of ward populations or patients considered at risk.

Anatomically, there are three venous channels in the calf which unite behind the knee to form a large vein which passes through the thigh to the groin, where a large tributary which drains the thigh joins. On each side, the large veins pass from the legs through the pelvis each getting another large tributary before they unite, at the back of the lower abdomen, to form the main vessel taking venous blood back to the heart. For an embolus to be large enough to threaten an otherwise fit patient's life it had to fracture somewhere in the single vein system. Any thrombus confined to one of the calf channels was not large enough if breaking off to cause death, passing on reaching the heart, out through the pulmonary artery to the periphery of the lung. By placing the transducer in four set positions between the thigh four inches below the groin and the lower abdomen it was possible to identify the site of any blockage.

It was now time to assess this method of investigation and compare it with the then current gold standards, namely radiology and radio isotope scanning.

First, it was established that in thirty normal volunteers venous flow could be detected from the calf to the abdomen. The examination was normal in all sixty legs. Next, I examined one hundred patients with possible deep vein thrombosis. These had either had a clinical diagnosis of thrombosis made or were strongly suspected of having had minor emboli which had passed through the heart to the periphery of the lungs where they were giving signs on examination.

The patients were either examined with Doppler ultrasound first or after ancillary tests but before the results of these were known. Apart from failing to up thrombus in the calf veins, where there are three channels, the correlation in the single channel veins, the site of potential fatal emboli, was excellent. Ninety-five percent of single vein thrombi and occlusions were identified when compared to the radiological studies. These findings once more established the inadequacies of clinical examination. Only fifty-five percent of legs thought to have thrombosis were confirmed as such, while in a further twenty percent of legs thought to be normal, significant and potentially life-threatening thrombus was found.

In this study it was shown that Doppler ultrasound was reliable as a diagnostic aid if thrombus extended into the potentially fatal zone. It would not detect more minor clots in the calf veins.

Since the start of my research, Frank Cockett had discussed the project with me on a weekly basis, giving me encouragement to keep at it, to continue searching the world literature and to continue down the avenues I was exploring. Like me, he became excited some six months after starting when the possibility of using Doppler ultrasound became apparent. It was then that the benefit of the work that I had done studying the size of emboli and their origin became important.

It was soon apparent that the findings on ultrasound examination correlated well with the "gold standard" radiological studies and were far superior to clinical examination. At this stage it was decided to publish a preliminary paper emphasizing that further studies were under way.

At about this time Frank, who was attending a meeting of Belgian surgeons in Bruges, entered a paper for me to deliver announcing these results. This was the first occasion and opportunity for me to talk at an international meeting. The weather was good and in addition to sitting in the square with fellow delegates, having a drink and listening to the carillon, we visited the famous "Old St. John's Hospital" one of the oldest surviving hospital buildings in Europe dating from around 1120. International surgical meetings were usually well attended and gave one the chance to meet other surgeons in a relaxed atmosphere and discuss mutual points of interest or research in which we were engaged.

On return, it was time to research whether Doppler ultrasound was suitable and reliable enough to be used as a screening tool to overcome the major shortcomings of clinical examination.

Because ultrasound examination was simple, rapid and reliable compared to other forms of investigation it appeared to lend itself to screening. To be useful as a clinical instrument, it needed to be available on a ward and to be suitable for use by ward staff rather than relying on a technician or doctor. Ten medical students volunteered to participate in this trial as investigators. Five were trained by me, in fifteen minutes, in the method of investigation and the other five were later trained by them. Twice a week, working in pairs they examined patients in four medical wards of thirty beds each, one female and three male, recording the results for each individual patient. The initial examinations were overseen by me to ensure consistency. Any abnormal results found during the survey were checked by myself and, where appropriate, subjected to confirmatory tests.

During the two month survey, one thousand seven hundred and seventeen patient examinations were made. The time taken for each ward survey of twenty five to thirty patients was just under one hour. During this time the first group of students identified more than ninety percent of patients with abnormal findings, while the second group trained in the technique by the first group only identified eighty percent. On studying these results, it was apparent that attention to detail particularly with regard to patient position was important. With this proviso the results showed that this test was far superior to clinical examination. It took about the same time to examine a patient as it did to take their temperature and thirty patients could be tested in the same time that one or, at the most, two patients could be investigated radiologically. With adequate training of ward staff it could be recommended that ultrasound screening was appropriate for everyday use.

I achieved relatively little during the first six months of research other than learning about the prevalence of death from pulmonary embolism, the dimensions of an embolus causing death and, despite great awareness of the problem, the inability of clinicians to diagnose the formation of clots in the legs on physical examination. It was only when I focussed on trying to make a diagnosis or spotting the clot in the leg, rather than attempting to retrieve an almost hopeless situation, that the task set to me by Frank became possible.

I was running out of time and Frank was eager that the work I was doing was completed, written up and published. It was necessary to apply for an extension to my research time and I was fortunate in that Sonicaid, realizing the potential of what I was doing, sponsored me for a further three months. This enabled the completion of the studies and gave time to put the work together and write my thesis for submission to London University for the Master of Surgery (MS) degree. Prior to the advent of computers, this was a lengthy task and required ten drafts before I and particularly my mentor Frank were satisfied with the way it was put together. Furthermore the English needed checking by an expert to ensure no grammatical howlers!

Once completed and six copies printed, it was submitted to the University. During the interlude while waiting for the next step, I was contacted by the BBC who recorded an interview and filmed an examination, prior to showing it on “Tomorrow’s World”, a scientific programme featuring developments to come. After two months, I was summoned for an interview at the Middlesex Hospital chaired by the Professor of Surgery, Leslie Le Quesne. Despite going with some trepidation, this turned out to be a pleasant occasion when I was made welcome and the thesis discussed. I was delighted a few days later to learn that I had been awarded the MS.

This was not the end of the project as Frank encouraged me to enter an abstract to the College of Surgeons and apply for a Hunterian Professorship, an additional honour and reward if appointed. An appointee was entitled to call himself a professor throughout the year. Once more I was fortunate and set about preparing my lecture. My reading of John Hunter's work and lectures led me to open my lecture with the following quotation

'Surgeons have been too much satisfied with considering the effect only; but in studying diseases we ought not only to understand the effect, as inflammation, suppuration, etc., but also the cause of the effect; for without this knowledge our practice must be very confined and often applied too late, as in many cases it will be necessary to prevent the effect.'

This extract on reflection appeared to summarize a lot of what the previous fifteen months had been about. I hoped the work reported would be a step forward in preventing the effect.

I duly delivered the lecture on the 13th.May 1971 in the Royal College of Surgeons before an audience of some two hundred, including family. Speaking in public was very much one of my weaker points but after a nervous start, which I well remember over forty years later, I then got into my stride and was able to deliver in some fifty minutes the substance of my topic with which I was well acquainted. The lecture appeared to be well received and many complimentary comments were made afterwards.

That evening Mary and I went out to dinner in Knightsbridge with my in-laws, a memorable and enjoyable occasion. In particular I remember my mother-in-law having prawns as a starter. A cascade of many prawns was brought to the table which she thoroughly enjoyed. There was something of a shock for my father-in-law when the bill came and he found that each prawn was individually priced!!

Other than attending one or two more conferences in Europe and reading a paper I also entered a paper for the Moynihan Prize at the annual meeting of the Association of Surgeons of Great Britain and Ireland held in 1971. Ten of the papers entered were selected. I should have had a good chance of winning that year but while the content was worthy, the delivery was poor whereas the delivery of the winner was excellent and the prize deservedly won.

My research time over I was now made a substantive Senior Registrar on the St. Thomas' Hospital training circuit. Before embracing this appointment, I was able to reflect on what I had gained from the previous fifteen months and how the benefits would aid my future career.

To get a consultant post in a desirable hospital it was necessary to have more to offer than being an FRCS. It had been somewhat reluctantly that I had set out on a research project. It was not long before I realized that the standard of papers written and published varied enormously in quality and content from one to five stars. I was soon able to discriminate between those worthy of further consideration and those bread and butter papers published to fulfil the annual expectations of a surgical department. Reading many erudite papers demonstrated the need to tackle a problem by going back to basics and accomplishing the ground work to have a structure on which to build. Thus, six months into my research, I realized that I had been looking at the effect and as John Hunter had pointed out it was necessary to look for the cause of the effect. By coincidence the timing of my research and the problem put to me coincided with advances in technology giving me an opportunity not previously available to others, a case of being in the right place at the right time.

Prior to starting my research I had thought that what I had learned as a medical student and subsequently would largely be set in stone for the rest of my surgical career. My time in research had taught me to come off the beaten path and to think laterally, coming at a problem from a different angle, a principle that has remained with me till this day. Furthermore the experience I had gained from reading so many papers enabled me on an informed basis to assess the content of new work being published. I had become fully aware that science and technology were moving forward at an ever increasing pace and that I should be on the lookout from that day onwards for any advances, which might be to the benefit of one's patients. This embedded philosophy was to surface again some years later.

Finally I learned from Frank the importance of overseeing and helping those who came to you for surgical training and ensuring that you could help them on the path to a successful career.

Chapter 8

The Final Years of Training

I was next appointed a senior registrar, the final step prior to being elevated to consultant. Each year one was rotated to a different firm to gain wider experience. I began on the firm where Kent Harrison, a thoracic surgeon, was the senior and Bert Thompson, a new appointee previously a senior registrar at St. Mary's, the junior consultant. Bert was an accomplished general surgeon and experienced at renal transplantation. He was appointed to set this up at St. Thomas' which had a high powered renal unit but, until his appointment' no means of transplanting kidneys in-house. He was sure of his abilities with almost a God-like personality in believing all he did would be fine.

Many patients were referred to Kent with lung cancer. Kent assessed the possibility of operating on lung cancer and other lung pathologies by entering the chest under general anaesthesia through a small vertical incision just to the left or right of the sternum, depending on which side the pathology requiring further elucidation was situated. In the case of lung cancer if the disease had spread to the lymph nodes or into the tissues around the heart, radical surgery to hopefully eradicate the disease was not possible. Patients recovered rapidly from this relatively pain-free operation and were fit to return home after three days. These patients would then be referred for radio and or chemotherapy to hopefully control the disease and possibly extend their lives.

If it was found that there was no involvement of these structures, the patient would be turned on its side and the chest opened, allowing sixty percent of patients the chance of cure following a radical removal of the diseased lung. Thoracotomy (opening the chest) was a painful operation requiring up to two weeks or more for recovery before discharge home. Those with inoperable disease were thus spared this procedure. When I arrived on the firm, the results of this minor invasion, which had been done by Kent for five years, were ready for analysis and publication so another paper was born which showed this method of assessment to be superior to others in use at that time.

A lot of patients were referred to Kent with heartburn, a most uncomfortable condition whereby acid in the stomach was regurgitated up the gullet. Untreated this could lead to inflammation and narrowing of the gullet giving increasing difficulty with swallowing. This came about because the hole in the diaphragm, through which the gullet passed to enter the abdomen, had enlarged allowing the sphincter at the lower end of the gullet and the adjacent part of the stomach to migrate into the chest. With the change in pressure on the other side of the diaphragm, the sphincter was held open allowing the flow of acid upwards and giving the symptoms of which the patient complained. In some cases, a patient could get a cancer in the diseased area due to long-term burning with acid

Where the gullet had not been shortened, due to the inflammation, Kent had devised a straightforward procedure to rectify this condition. The upper abdomen would be opened and the stomach drawn back down below the diaphragm and then fixed with sutures to the under surface of this to prevent the sphincter again riding up into the chest. With the correct patient selection this proved a highly successful operation, giving instant relief of symptoms and a good long-term result. This was yet another treatment that I took on board and used down the years.

On the other side of the firm Bert set about organizing for in-house renal transplants while at the same time performing complicated general surgical procedures to a high standard. Among the more complicated was oesophagectomy (removal of the gullet) for cancer. This could be a two or three stage operation. In both, the abdomen was first opened and the proximal part of the stomach mobilized so that from the gullet to two thirds down the stomach all connections with other tissues were separated. The viability of the stomach was maintained by preserving the blood supply at the distal end and at the same time carrying out a manoeuvre to allow drainage of the stomach into the small bowel. The abdomen was then closed and the patient turned on its side and the right chest opened.

If the cancer was confined to the lower third of the gullet, the most common site, the lower half of gullet would be mobilized, again all connections with surrounding tissues being divided. The stomach would then be drawn up into the chest and the diseased lower half of the gullet excised. The stomach would then be anastomosed (joined) to the upper half of the gullet and the operation completed.

In those patients with disease in the mid or upper gullet, the chest would again be opened and in this case the gullet mobilized as high as possible in the chest. Following this the chest was closed and the patient turned again on to its back to enable an incision to be made in the neck. The upper part of the gullet would be identified and mobilized. The gullet would then be delivered into the neck drawing the stomach up which would then be joined to the upper gullet at a suitable point. This was a very major operation for the surgeon but particularly the patient the whole procedure taking nigh on five hours. Until healing had taken place, the patient would be fed through a tube passed through the nose down into the stomach.

Patients requiring a renal transplant would have their tissues analysed and typed. A central register of recipient tissue types was kept in Cardiff. When a potential donor became available, more often than not being brain dead as result of a road accident, the donor was tissue typed. The results would be correlated in Cardiff, whereupon the centre would notify a transplant hospital that they had a match for a particular patient. The relatives of the brain-dead victim would be interviewed, usually a harrowing experience, and permission sought to use tissues for transplantation. In most cases, once the situation was explained, permission was given.

As the senior registrar to Bert, I was trained by him to remove a kidney from a donor so that it could be used in the recipient. At that time it was important after removing the organ to cool it quickly, as those with a short warm time after removal did best and functioned more rapidly after transplantation.

The typical scenario would be that we would receive a call from Cardiff to say there was a donor for one of our patients. If it came from some distant part of the country it would be removed by a locally trained surgeon, bagged, cooled down and flown in a private plane to our nearest suitable airport. From there a car would pick it up and deliver it. If the donor was within our area, not more than an hour and a half away by road, I would set off in the car to the hospital more often than not in the late evening so as not to disrupt normal hospital schedules. On occasions, my wife would accompany me. On reaching the hospital and having checked that all the requirements regarding consent and the donor's condition were satisfactory, the patient would be brought to theatre when the life support equipment would be turned off. We then waited for the heart to stop before harvesting began. The organ would be retrieved, cooled and placed when bagged in a bucket containing ice. I would notify Bert that I was setting off and he would then organize for the recipient to be taken to theatre and prepared for operation. Once in the car, on the return to St. Thomas', my wife when with me would support the bucket between her feet, speed being of the essence I did not linger on the journey.

Meanwhile the recipient, who like all others on the transplant waiting list would have been on standby for this moment, would be contacted, rushed into hospital and prepared for theatre. On arrival of the kidney in theatre, the organ would be checked to confirm its suitability and the patient anaesthetised. The lower abdomen would be opened and the anatomy defined. The artery and vein on the kidney would then be anastomosed to the suitable vessels in the recipient's pelvis and the ureter (the pipe from the kidney to the bladder) joined to the bladder. On completion of the operation, the recipient's body would be immuno-supressed and the patient taken to a side ward to be barrier nursed until the drugs could be reduced and the kidney started functioning. This could take a few days, to some extent depending on the warm and cool times of the donor organ prior to re-establishing circulation in the recipient. This was in the early days of transplantation, prior to the development of improved means to preserve the donor organ out of the body, and was quite dramatic. With a successful transplantation, the recipient's life was transformed, as the patient was no longer dependent on dialysis several times a week to maintain life.

For an insight into Bert's character, one Friday we operated on a patient doing a two stage gullet resection. All went well in theatre and the patient was transferred to the ITU (intensive care unit) for recovery. Next morning before departing with Mary to see my parents, now both retired, for the weekend I visited the ITU to check on the condition of the patient. I was somewhat alarmed by the X-ray taken that morning which showed air in the right side of the chest. I rang Bert and told him there was a problem. He answered over the phone that this could not be so, as it couldn't happen to one of his patients. I stuck to my ground whereupon he got quite angry, saying he would come in himself and I had better not be there when he arrived. I replied I had no intention of waiting around as I was already late in departing for the country. He duly arrived and after confirming my observations were correct, did what I had suggested. On a ward round with a number of others, we next saw this patient together, now recovering well. He looked up at me and just said "This patient has done very well". The patient went on to make a good recovery prior to being discharged home. The year passed quickly and once more it was time to move on. For the future, I had learned how to approach and treat patients with oesophageal pathology and how to approach relatives of patients proven to be brain-dead to request the donation of kidneys and occasionally other organs. When faced with a brain-dead patient under my care, I would think of the possibility of being able to add to the somewhat meagre supply for those awaiting an organ transplant.

Since our marriage, Mary had completed her training and then was appointed to a charge nurse post in the Radiotherapy department at St. Thomas'. This enabled her to work much more social hours. After about a year she applied for and got a post as a nurse assistant in a pathology clinic in Harley Street. This involved collecting specimens and blood samples from patients referred to the clinic. During the course of this, she met many interesting people. With both of us working, we would often in the evenings cross the river and go for a meal in Chelsea or Fulham. In 1972, we had our first child, Katie, and I was able to be present at her birth, an exciting and happy moment for us both. Luckily we had a first floor flat so Mary was able to manage getting out and going to the shops or for a walk over the road in Battersea Park. Her mother was a great help to her at this time.

After an enjoyable first year as Senior Registrar, I began to look to the future. At that time, appointees to consultant posts averaged fourteen years in training before attaining this status. I began to apply and attended a number of interviews but mostly it appeared that the successful candidate had been sought after, or else it was the most senior trainee's turn. About this time, my friend John Fergus, now a consultant urologist at Northampton, and with whom I had shared a flat earlier, got in touch with me and asked if I would like to do a locum consultant post, as the senior consultant had had a heart attack and was to be out of action for a number of months.

After discussion with my supporters at St. Thomas' I took on the post. It gave me some valuable experience into what life was like on the other side of the fence, which was to enable me to hit the ground running when eventually appointed. We were given hospital accommodation which turned out to be the end building of a modern terraced block of houses with small cramped rooms. It was poorly constructed, cold in the spring and overheated in the early summer. In addition, the walls were thin and pervious to sound from next door. Mary was pregnant with our second child and it was therefore not a happy experience for her. At thirty-seven weeks, she returned to our flat in London while I commuted as and when my duties allowed. On one of these occasions we went out to dinner, leaving her Mother in the flat to baby-sit. I returned to Northampton that evening and became violently sick with food poisoning. While in the middle of this, Mary phoned to say she had gone into labour and was on her way to the Maternity Unit at St. Thomas'. I was in no fit state to jump in the car and return to London, so missed the birth of our son. Happily all progressed well and I was able to collect her from the hospital to return to our flat. It was a lovely Spring day and we stopped at a pub in Chelsea on the way back. After a time we got up to go and I was halfway back to the car before I remembered I had left our son in his basket under the table!!

We spent another couple of months in Northampton before the Consultant on sick leave was passed fit to work and we returned to

London. My absence had been covered by a locum Senior Registrar and I joined the firm run by Frank Cockett and John Pullan.

John was a great clinician whose surgery was a joy to behold: neat, precise and dextrous. His speciality was endocrine surgery involving surgery to the thyroid and parathyroid glands in the neck and the adrenal glands sitting above both kidneys. John had a large tertiary private practice and it was not long before I discussed this with him. His secret he said was to listen carefully to what the referred patients told him and then take a detailed history. In over half the cases referred to him this had not previously been done. Following clinical examination, he was able to translate the history and findings, in around sixty percent of patients referred, into a clinical diagnosis. For the remainder, he was unable to offer anything more.

When operating in the neck, John would make an incision usually half the length of that made by other surgeons and well down, so that the scar would give a good cosmetic effect and could easily be covered by a necklace or choker. In his hands the anatomy would be well displayed in a bloodless field and the resection of thyroid tissue or tumour neatly performed. Operations on the adrenal glands situated below the diaphragm and above the kidneys were most commonly done for Phaeochromocytomas (tumours of the adrenal, usually unilateral). The hormones put out by these tumours could cause sweating, palpitations and elevation of the blood pressure to quite alarming levels. Patients undergoing surgery for the removal of one of these, required careful pre-operative preparation to block these effects as handling during removal could, before the blood supply to the gland was secured, give a surge in hormone output. This could give embarrassing side effects such as a huge surge in the patient's blood pressure which the anaesthetist would have to rapidly counter. As this moment in the operation was approached, it would give rise to a few tense moments. The surgeon needed to be dextrous and delicate with his handling.

Having previously worked with Frank and knowing my capabilities, he decided to take a year off from vascular surgery. He would do general surgery for the year and I would do the vascular. There were between fifty and sixty patients on the waiting list and this figure had been constant for some time as the moment one patient came off, another would be added. The surgery encompassed all vascular surgery below the diaphragm down to the upper calf. There were patients with aneurysms while most had blocked arteries due to arteriosclerosis, mostly with a history of smoking or, as an associated disease, diabetes.

Doing lists with two to three vascular cases I managed during the year to get the number of patients on the waiting list down to around fifteen. Frank humorously thought there would be no work for him to do at the end of this time. By now I had a set routine for performing operations and the highly competent theatre nursing staff took advantage of this to train future theatre nurses. The trainee would be assisted and overseen, usually for her first few cases, by a charge nurse or sister until it was felt that the 'L' plates could come off. The standard of assistance achieved by the nursing staff was second to none and was highly important to the smooth running of what could at times be difficult operations. This reinforced that while the surgeon was in command for each operation nevertheless each procedure was very much a team effort and the outcome particularly in difficult cases, depended not only on the surgeon, but also his assistant, the anaesthetist, the theatre sister and any technicians.

The system at St. Thomas' had now changed in that the post of Resident Assistant Surgeon was no longer a one or two year living in appointment. There was a rota whereby a trainee, usually the senior, would live in on the day or weekend that his firm was on duty. On one of these occasions a man of sixty-eight was admitted in a collapsed state with backache, and abdominal pain. On examination he clinically had a ruptured aortic aneurysm, one of the lucky ones surviving long enough to reach hospital. Bloods were immediately taken and an X-ray done in Casualty which confirmed my clinical diagnosis. The emergency buttons were pressed, and my adrenaline began circulating. The patient was rushed to theatre where the team was already assembled and preparing for action. While getting ready I had a quick word with the anaesthetist to ensure that no paralysing agents were given until all was ready to put knife to skin.

The patient, lightly anaesthetised, and whose general condition remained stable, was placed on the operating table and skin preparation and towelling up done. I quickly discussed with the theatre sister the order in which I would want various instruments and retractors and despite the tension of the moment I felt icily calm. I began by making a long vertical skin incision in the middle of the abdomen, the anaesthetist reporting all was remaining stable. It was now time for the patient to be paralysed and as the drugs were working I rapidly entered the abdomen and placed the retractors to give access. The gut was then brought to the exterior and placed in a bag. I could now see the large collection of blood behind the peritoneum, the posterior wall of the abdominal cavity. This patient had been fortunate in that the rupture was in this position as if it had occurred anteriorly into the tummy cavity, death would have rapidly occurred. I now incised through the posterior wall and managed with a mixture of digital and instrument dissection to get round the aorta above the aneurysm and apply a clamp.

From skin incision to this point had taken some five minutes and, with the clamp applied, we could all relax as the situation was now under control. An air of calmness settled and the operation now became routine. The aorta was fully exposed and further clamps applied distally to enable the aneurysm to be opened. A Dacron graft was inserted and stitched into place utilizing techniques I had learned from both Peter Martin and Frank. On release of the clamps, blood flow to the lower limbs was restored and the presence of pulses in both feet confirmed, indicating that no debris or possibly displaced clot had embolised distally blocking the arterial circulation. At the end of the operation the patient was transferred to the ITU for recovery. On a ward round with Frank a few days later he informed me that this was the first patient to survive surgery for a leaking abdominal aneurysm at St. Thomas'. Some days later the patient was discharged from hospital and was well when seen at follow up a few weeks later.

On another occasion, a male patient of some seventy years was admitted with a sub acute large bowel obstruction. After preparation, he was taken to theatre and his abdomen opened. He had a cancer of the large bowel which was operable, meaning that with excision the prognosis was good. After a somewhat lengthy and difficult operation with resection of the diseased area and reconstitution of bowel continuity, the procedure was completed. The patient was a Jew and had a very concerned and caring family. They were informed that the outlook was good and that he should make a good recovery from the operation.

However, the patient had other ideas and decided that he wouldn't get better. Many times in my clinical career I found that if a patient decided to turn his or her toes up, there was little one could do to change the outcome. Similarly patients you might not have expected to survive, by dint of will power went on to recover. I spoke again to the relatives who encouraged me to do all in my power to pull the father through. With this stimulation, I decided to throw the book at him. I abused him by saying things such as after Auschwitz, he should have more respect for life. I told him and reinforced the predictions that he had a good outlook if he made the effort, and that I was not going to sit around and let him slip away. He was also told that he appeared to have no respect for the time and skill which had been devoted to his care. After about three days of bombarding and abusing him, I suddenly noticed a twinkle in his eye and a change in attitude. From that moment onwards, I was able to tone down my attack and he went on to make a somewhat lengthier recovery than usual, probably due to his intent to die for the first six days after operation. Whenever he saw me, once the corner was turned he gave me a welcoming but possibly guilty smile. The family were delighted with the outcome and on his discharge from hospital presented me with a case of wine. I had learned another lesson. Recovery from a major operation in addition to the physical element also had a mental element which, in this case, the treatment of the latter had been taken to the extreme.

I was now fourteen years after qualification and realized it would soon be my turn to be appointed a consultant. Frank indicated that if I wished I could probably get an appointment as a vascular surgeon in London. He also intimated that, if I looked outside, I should ensure that any hospital had all the back up facilities in the way of radiology and pathology that I would require. I applied for a number of posts tweaking my CV on each occasion to fit the requirements of the post. Mary and I discussed where we would like to be. I had been brought up in the country and by this time had spent more than twenty years in London.

The life of a London consultant such as Frank seemed to consist of visiting private patients at nursing homes where you had operated at the beginning of the day, then attending your teaching hospital to carry out medical work, teaching, administration, and supervising any research projects. On leaving your hospital you might go on to perform some further private operations or attend some meeting of your speciality at the College of Surgeons or the Royal Society of Medicine. On top of this, travel in London was not easy. If I travelled from my flat to St.Thomas' at a quiet time the journey would take ten to fifteen minutes whereas in rush hour, to be avoided at all costs, it could take seventy five. During my time on this firm I had on a number of occasions gone with Frank or John to assist them with operations at the London Clinic, King Edward VIIth, the Royal Masonic and various Nursing Homes. This further underlined the difficulties of getting from A to B. With such a full time career, this would have left little time for family life. While I fully appreciated that the demands of work would be much the same, nevertheless it was likely that any travel time would be much curtailed.

I did not apply therefore for a London post.

Chapter 9

Consultant Appointment

Throughout 1973, I was searching earnestly for a consultant post. I looked at a number of possibilities and sent in a few applications, on each occasion having to tailor my CV, to fulfil the requirements of the advertised job description and emphasizing the aspect for which they were looking. When a possibility came up, it was necessary to meet the surgical staff of the hospital concerned, view the facilities in particular the operating theatres and the wards where, if appointed, one's patients would be nursed. I would also check on the support disciplines, mainly pathology and radiology, and determine if possible whether the family would enjoy living in the area for many years to come.

After a number of disappointments, I was short-listed for a post in Shrewsbury to replace the leading and most successful surgeon who was retiring. After my preliminary visit, I was short-listed and recalled with the others on the list to meet members of the staff so that applicants could be assessed. At that time there were five on the surgical staff, one being a urologist, treating a catchment area numbering over three hundred thousand. The geographical boundaries extended from a point midway between Telford and Wolverhampton in the East, to a point well into Wales mid way between Shrewsbury and Aberystywth in the West. To the North, Whitchurch was the upper boundary, while to the South patients came from beyond Ludlow where the boundary from there extended to the West and included what was previously Radnorshire, a distance of some forty-five miles from Shrewsbury.

On meeting members of the staff, I found that a number were London trained and like myself had wanted to move out of the big city. Their quality and enthusiasm was stimulating. I also met the chairman of the Area Health Authority, Frank Leith, who was to sit as the lay member on the interview committee.

The interview took place in the autumn at the Regional Health Authority building in Birmingham and among those present on the committee were the next senior surgeon and the lay member from Shrewsbury, a College of Surgeons' representative to ensure the proposed appointee was suitably qualified, a Surgeon representing the Medical School in Birmingham University and a lay Chairman. The candidates were interviewed in alphabetical order so I did not have long to wait. The interviews of the five candidates each took some twenty-five minutes during which time probing questions were made into our experience, what we had to offer for the future and what attracted us to Shrewsbury. Deliberations after the last interview went on for what seemed ages before I was summoned back in and offered the post which I readily accepted. The years of training following qualification and my age at the time of my appointment were the exact average for promotion to consultant level. All the surgeons at Shrewsbury had trained outside the Region, and I subsequently learned that the deliberations had been lengthy because Birmingham wished to appoint one of their own. Luckily his CV did not compare with mine, so the locals won the day.

News of my success was welcomed at St. Thomas' where, having received confirmation of my appointment from the Regional Health Authority, I gave in my notice and set about making plans for our move to Shrewsbury.

I was to take up the post at the beginning of March. A number of visits were made when Mary and I met future colleagues and set about finding somewhere to live. Because of distances involved we decided to rent initially and look for somewhere permanent when in post when we would have a better idea of the local geography.

After winding down and saying fond farewells, we left St. Thomas' where I had studied and worked on and off for over twenty years. While there I had become a founder member of the Cheselden Club, named after a prominent Thomas' surgeon of the eighteenth century. This was an internal surgical society which met once a year. Papers covering clinical matters such as case reports and research projects would be presented. Pursuant to the afternoon session, a paper would be delivered by an eminent guest speaker prominent in the surgical world followed by a dinner. All in all these proved instructive and enjoyable occasions.

We had sold our flat in Battersea for a good price having been fortunate in being able to buy it originally as sitting tenants. By now we had two cars and set out independently for Shrewsbury on February fourteenth.

On the journey, I mused over what I had learned over the years and what I would need to put into practice in the years to come. First I had to remember that I would always be treating a patient and not a condition and secondly I should always listen to patients when describing their symptoms and be alert for any give away remark which might indicate their true concern. While a letter from the patient's doctor would always come with them the conclusions drawn could be wrong. On occasions, a remark from a relative accompanying the patient again could be most helpful. When in hospital under my care I would be responsible at all times, when available, for the patients' care and it was paramount for me to ensure that this was understood by my junior staff in training. The buck stops here.

With regard to the condition to be treated I realized that I had had a comprehensive training, covering and having considerable experience in gastro intestinal surgery, vascular surgery, urology, endocrinology and thoracic and paediatric surgery. I had worked and been trained by some of the best surgeons in the land and had been fortunate to work in some of the best hospitals. It was now down to me.

At the journey's end, I pulled up at the property we were renting on the Mount, a short walking distance from a house once owned by Charles Darwin. During the next two weeks we set up house, retrieved the children from my wife's family and explored Shrewsbury seeking the necessary establishments to ensure order in our lives. On one occasion exploring the shops, my wife heard a foreign language being spoken and asked if I knew what it was. I recognized it immediately as Welsh, a reminder that we were living on the English/Welsh border. During one of these expeditions, I remarked to my wife that this would be the only time that we would be incognito in our new surrounds.

On Friday first March I attended the hospital, signing myself in and meeting some of the clinical staff. I met my predecessor on the following day when he took me round what had been his wards but were now mine. We met the nursing staff before we went together, he doing his last and I my first ward round. He introduced me to the patients and handed them over to me with a résumé of their surgical state. After our round Jack Baty left the hospital never again to return to the wards.

The Shrewsbury Hospitals at this time were on three sites. The Royal Salop Infirmary (RSI) situated in the middle of the town had been established early in the eighteenth century. Half the medical and surgical services were based here together with the outpatient departments where clinics were held. To the west going out of the town was Copthorne Hospital on the left of Mytton Oak road, in the main a hutted hospital built at the time of WW1. This had ten wards, Nightingale style each containing thirty beds, arising off a main corridor together with operating theatres and an X-ray department. Situated here, there were general surgical, medical, paediatric, gynaecological and short stay facilities. Across the road the Royal Shrewsbury Hospital was building and taking shape. At this time only the maternity and pathological departments were functional.

On the Monday, I did my first ward round with my inherited junior staff, an Australian registrar and House officer. The ward sisters were both excellent, one having been in post for many years, down to earth with good knowledge and judgement, happy to roll her sleeves up and not afraid to express her opinion. The other was London trained and fully aware of what I was about. I impressed upon them that we were a team looking after those entrusted to our care.

The surgical consultants worked as two firms each having a trainee Registrar but having our own house officers. I was paired with a Scotsman, David Pollock, who trained in Glasgow and was similar in age to myself. When one of us was away, the other covered all the patients on the firm. We worked a one in four on call emergency rota, Mondays to Thursdays being fixed with the weekends being covered in turn. When on call I requested my house officer to phone me at 10.00pm. This gave the incumbent valuable experience in communicating clinical data and honing their skills to impart essential information. It updated me on the emergency admissions and filled me in on their management. If I sensed any problem or difficulty in diagnosis I would return to the hospital, a five minute journey, to assess and depending on the experience of the registrar either leave him to care for the patient or go to the theatre myself. If the registrar was going to operate on anything out of the ordinary, I requested that he would ring and discuss with me before going to theatre. If the operation was within his capability, I would ask him to proceed but to call me if he had a problem or would go into the hospital and either mentor the trainee or operate myself.

Following the ward round, we would have a coffee when I could discuss any interesting clinical situation we had seen. This was a habit I first picked up at the Hammersmith some years earlier and proved a good teaching tool in addition to planning the future management of any individual patient, particularly those who had been admitted as an emergency.

After the Monday round, I went to the Infirmary to the Outpatient department to see new patients. The Infirmary was a bit of a warren continuing to function until the new hospital was commissioned. The outpatient secretary there was to move shortly to Copthorne where she became mine and the firm's secretary for some twenty-five years. This was a lucky break as she proved a gem and could always find me even before the days of mobile phones and would ensure, as she kept my working diary, that I was in the right place at the right time. She was also able to determine from speaking to the GP the degree of urgency for a visit from me. General Practitioners knew to contact her if they needed my services.

It was not long before GPs began requesting that I did a domiciliary visit to see patients in their home. Often these would be done at the end of the day before going home. These had a number of benefits. First, I could meet the doctor and we could assess each other and then I could discuss a complicated case with the GP. For the patient, in the case of possible cancer or other life or limb threatening condition, it had the advantage of short circuiting access to specialist treatment enabling, where necessary, earlier admission to hospital. Secondly, it gave me the ability to determine the degree of urgency in getting a hospital admission. Thirdly, in elderly patients, particularly those requiring major operations, I could see whether the home was suitable for early discharge from hospital or whether, before returning home, the patient would be better cared for by convalescing for a while in a Cottage hospital. I well remember visiting a house in Dawley, now part of Telford, which still had lamp lights there being no electricity fitted and a privy at the end of the garden, an abode not suitable for this patient for early return from hospital.

On the Tuesday, I did my first operating list the patients having been selected by the Registrar. This proved a light list but gave me the opportunity to meet the senior sister who had trained in London and was married to a gastroenterologist, one of the physicians. We were able to discuss how I liked to work and in particular any instruments I might require. The anaesthetist was a Scotsman, shrewd and very competent for routine general surgery. His wife ran a practice in Shrewsbury. These were the days before extensive monitoring in theatre and patients for major procedures would have just the blood pressure and pulse rate recorded. Jimmy would have the next patient on the table almost before I had been able to complete the notes of the previous operation. He became a friend and father figure giving me sage advice. Within a month of starting his wife had referred me my first private patient which I took as a compliment to my abilities and skills. On completion of the day's operating, I reviewed the patients on the waiting list finding that it extended to some six hundred names. What a challenge!

On Wednesday I journeyed in the morning to Telford to do an Outpatient Clinic at Wrekin Hospital (shut down in the 1980s after the opening of the Princess Royal Hospital) followed by a then free session available in due course for private practice.

On the Thursday, I had another full day of operating beginning at 9am. Subsequently these lists often went on to 6.00pm plus. I was very fortunate to have a recently appointed anaesthetist for these sessions who had trained at St. Mary's hospital in London and had worked there on the vascular unit run by Felix Eastcott, another of the top five vascular surgeons in the country. When I came to operate on major vascular cases this was a bonus as more refined monitoring of the patients' wellbeing and experience in the management of fluctuations of blood pressure were vital for the patients' survival and the success of the operation.

On the Friday, I did a further ward round and reviewed in particular those patients operated on the previous day before going to the RSI for a follow up session. There I reviewed those patients who had recently been in hospital for treatment and patients on longer term follow up for conditions such as cancer or vascular insufficiency usually associated with hardening of the arteries or diabetes. For efficiency, I stopped my secretary coming to outpatients and used a Dictaphone, a time saving device all round. After the morning clinic the afternoon was free, in time to be used for private consultations.

Shortly before I arrived in Shrewsbury, a Nuffield Private Hospital less than ten minutes from the out of town hospitals had been built and opened, one of the prime movers to get this off the ground being my predecessor. This had good outpatient facilities, rooms for thirty in-patient beds in two wards, an X-ray department and two operating theatres with a good recovery area adjacent.

In addition to these commitments, once a month on a Monday afternoon, I went to the world renowned Robert Jones and Agnes Hunt Orthopaedic Hospital in Oswestry to do another Outpatient session mainly for the population living in that area but also to attend any in-patient in that hospital requiring a general surgical opinion.

Although bleepers were in use in the hospital, this was the era before mobile phones so when on call for emergencies one's movements and position needed to be known. When not in the hospital, one could not venture far from a phone. Serious road traffic accidents giving severe injuries and dire vascular problems, such as a ruptured aortic aneurysm, needed to be seen and assessed rapidly to enable treatment and improve the chances of survival.

After the first week I began, in particular, to try and sort the waiting list out. In the main wards, which I shared with my colleague I had fifteen male and fifteen female beds. Of these we aimed to have two beds for each sex empty and available for emergencies when going on-call. There was also a five day ward open Monday to Friday which took short-stay patients for minor and intermediate procedures and also transfers from the main ward of patients being due to be discharged before the weekend. Keeping within the code I aimed to use the main beds to the full and fit as many short-stay patients in as possible to ensure full operating lists. This meant operating on four or five major cases per list and aiming to make the number per list up to fourteen maximum with minor and intermediate cases. This was fine in principle but the number of emergency admissions could be very variable and inevitably lead to overflow into my colleague's beds on more than acceptable occasions. Understandably there was an element of aggravation at times when due to lack of beds he could not admit patients for operation. I suppose I was influenced wrongly, knowing I started with six hundred plus on the waiting list while he, having started with a clean sheet, only had some two hundred plus patients. The waiting time for removal of a minor skin lesion for cosmetic reasons or a subcutaneous fatty lump at this stage could extend to more than five years. All the time patients with cancer or other life or limb threatening conditions were given top priority. Others with excruciating recurrent attacks of pain such as occurred with gall or kidney stones which severely interfered with their lives and increased demands on the GPs were also prioritized. From this it will be apparent that I had to run to have any chance of giving the semblance of a service to the patients.

This was to be the pattern of work for the next few years, pending the appointment of further surgical staff.

Within a few months we purchased a house two doors away from our rented property and moved in. This had more rooms and allowed me to have a study/consulting room. On a fine day, the family could cross over a stile a few yards away, descend down a field and walk along the grassy banks of the river Severn.

Chapter 10

Getting established

After the first week, life settled into this pattern with a vast volume of work.

I rapidly became aware how fortunate I was with the support-medical disciplines without which it would have been difficult to establish a sound clinical practice. We had excellent radiologists one of whom, Michael Dean, developed into a vascular interventionist of national and international renown. At the time of my arrival he was investigating peripheral vascular disease by inserting a catheter in the groin into the femoral artery and threading it up into the aorta before injecting contrast media. He rapidly took on the technique of inserting a needle through the loin into the aorta and injecting radio opaque dye, thus obtaining much improved definition of the arterial tree from the aorta to the lower legs. This enabled better assessment of any disease and improved guidance as to the most appropriate surgical procedure. Over the coming fifteen years he honed and advanced his skills to such an extent that interventional radiology became a robust alternative to open surgery.

The microbiologist was of similar age to myself and had trained in King's Hospital in London. He was a great protagonist for taking specimens for culture and only using a very limited range of antibiotics. If indicated by the patient's condition and drug sensitivities when available, he would only then use a new antibiotic. He was not confined to the laboratory and visited the wards to give specific advice where needed. This principle I had favoured for years after my experiences at St. Peters Hospital, Chertsey. Following this regimen antibiotic resistant bacteria were rare.

The histologists were both excellent and would refer difficult specimens, where there was some doubt as to their precise nature, to national panels, one of them being a member of one of these bodies.

This helped to ensure that once more patients were offered the correct and best treatment. Finally there was also excellent biochemical support especially in complicated situations.

The Paediatric Unit was also based at Copthorne and was run by Jack McCauley and Peter Capps. Both had trained in internationally recognized centres in London. As the hospital was close to Wales, there was a requirement for nurses to be Welsh speaking as English was often not the first language of those coming from across the border. My fellow consultant on the firm, a David like myself, had also had training in Paediatric surgery. Paediatric trainees when needing to refer a child for surgery would phone Jack for advice. The standard answer soon became, I don't mind who you refer to providing his first name is David.

During the next few months, I worked steadily at the waiting list covering the full range of general surgery, including upper gastro-intestinal, colorectal, genito-urinary, endocrine and paediatric surgery. At first, there was little vascular surgery as my predecessor had not taken on this relatively new field, the work in the main being done by my surgical colleague. However, the catchment area population being the size it was, there was far too much work for any one individual so this field steadily expanded.

One day, I was presented with a patient in her forties with a mass in the right groin. She was admitted to hospital and a biopsy confirmed her to have a lymphoma, a malignant tumour. A radical removal of the lymph nodes in the right groin region was done. This was followed by radiotherapy to the region. She went on to make a good recovery from this and I followed her up as an outpatient. Over the course of time, she developed a mass in the right loin region adjacent to the right kidney. A needle biopsy, on histology, confirmed a recurrence of her lymphoma. Despite further therapy, the mass being inoperable continued to enlarge and over the course of one year reached the size of an adult head and was causing her physical distress. She was a very stoical and somewhat laid-back person. She could see that I was concerned about her condition. I was unable to offer any further treatment, none being available. She smiled and in an attempt to give her some hope I gave her a further appointment to come and see me again in three months. Once she had left the room, I rang her GP and discussed her situation with him. He had been seeing her regularly and was well aware of her rapid deterioration. We both considered that she was almost at the point of needing terminal care and that she would not survive the month.

Three months later her name was on my list to review in outpatients. I expected to have an empty slot. Much to my surprise she entered to see me.

I noticed immediately that although still distressed, her condition, in the intervening period, had improved. On examination the abdomen was still distended and the mass easily palpable although it appeared a little smaller. X-rays and scans showed no evidence of the tumour having spread elsewhere. There being no further treatment to offer I confirmed to her that clinically there was some improvement in her status and again arranged to review.

Over the course of the next few years the tumour mass regressed to become impalpable on clinical examination. Serial scans during this period confirmed shrinkage of the tumour to leave a small scarred mass in the region of the kidney. As time went by, I was reminded of Chernobyl where the disaster had been confined by entombing the nuclear hazard in concrete. In this case the patient's own tissues performed this effect. I continued to see this patient over twenty years when all remained stable. She continued to smile and I came to realize that I had witnessed a real life miracle, the patient in this case not having been to Lourdes.

On another occasion, when on duty for the weekend in my early years, I received a call from a GP in Bishops Castle, South Shropshire. He asked me to see a man with severe abdominal pain. He arrived at the Nuffield Hospital and when I saw him appeared to give the impression that he didn't know what all the fuss was about. He was a man of the country and organized shoot parties and venues. He told me he was an accomplished shot and was in the Guinness Book of Records for having shot the most pigeon in a day. On questioning about this he said he had built a hide and set up decoys. His bag for the day numbered over five hundred.

On questioning him about the reason for his attendance, he said that he had developed lower abdominal pain a few days earlier but that this had got considerably worse overnight causing his wife to send for the doctor. On examination he had generalized peritonitis and I recommended immediate hospital admission with a view to operation. He looked at me and said he couldn't possibly come into hospital as he had shooting commitments but could be admitted a few days later. I knew he was sitting on a time bomb and said to him if he did that, I wouldn't be operating on him but attending his funeral. His wife then applied pressure and reluctantly he was admitted. Later that day I operated on him for a perforation of the distal large bowel. This required a two stage procedure from which he went on to make a virtually uneventful recovery. When he came to see me as a follow up some time later, he said that I was the first person in his life to tell him what to do and get away with it. I replied that at least he was there to tell me about it. As a hobby, I had taken up shooting to get some fresh air and exercise during the winter and avoid SAD (seasonal affective disorder or winter depression); a mood of depression seeming to pervade my colleagues in the hospital during January and February before the arrival of Spring. My patient remained grateful for what I had done and I became in subsequent years his guest at shoots on a number of occasions. It was a privilege to stand in the line and appreciate the skills of some outstanding guns.

It was not long before when on call, my registrar rang to say that he had just seen a patient who he thought could have a ruptured abdominal aortic aneurysm. I asked him to notify theatre, and the X-ray department and to order blood. Within a few minutes I was in the hospital and agreed with the clinical findings. While awaiting an X-ray to confirm the diagnosis, I ensured that all was under way to take the patient, a man in his late sixties, to theatre. I contacted the anaesthetist, who as luck would have it both for the patient and me, was the London trained contemporary with vascular experience. On learning of the nature of the case, the theatre staff had summoned an on call senior sister, again with some vascular experience. I discussed with her what I would be doing and what instruments I would require until I had managed to get clamps on the aorta and have control of the situation.

While I and my assistants were scrubbing up, the patient was brought into theatre conscious and placed on the operating table. The abdomen was then laid bare and once my team and the theatre sister were ready, I then began to prep the skin with iodine and towel up. During this time the anaesthetist was recording the patient's vital statistics and reported that the blood pressure was remaining steady at 90-95. With the towels in place, the patient was put lightly to sleep and I incised the skin through a vertical incision from the xiphisternum to the lower abdomen and down to the underlying muscle layer. It was now that speed became important. A muscle paralysing agent was administered and as the muscles relaxed I divided them and entered the abdominal cavity. I confirmed the presence of a leaking aneurysm, the blood being contained behind the peritoneum, the posterior wall of the cavity. Retractors were placed and the abdominal wall held open to allow access. The abdominal contents were rapidly mobilized and brought to the exterior and placed in a transparent bag. The anatomy of the posterior abdomen was now fully and clearly visible. The posterior wall was next opened and with a mixture of finger and instrument dissection I got round the upper aorta and applied a clamp. Despite the urgency I had remained almost icily calm throughout.

The heat of the moment was now off and an air of calm descended. My pulse rate was able to return to normal. Below the division of the aorta into the vessels to supply both legs I secured distally the iliac vessels which were beyond the site of rupture. I was now able to take full stock of the situation. While doing this, the anaesthetist replaced some of the blood into the circulation restoring the blood pressure to a more normal level. In this patient the rupture being in a favourable position I was able to insert and suture in place a straight Dacron tube graft to repair the defect. This was done without undue difficulty. It was then time to ensure that there were no leaks from the joins before slowly reopening the circulation to the lower limbs. This had to done slowly while the anaesthetist increased the rate of blood transfusion to maintain the blood pressure and ensure no catastrophic drops which might endanger the patient's chance of recovery. With this under control and the clamps fully off, the feet were examined to ensure the presence of pulses and that no debris had gone distally to occlude the circulation. With all satisfactory the abdominal contents were replaced and the operation completed. The patient was then placed in the recovery area, there being no ITU at this time and monitored until the morning before transfer to a side room in the main surgical ward. He went on to make a good recovery and was able to be discharged home after ten days. When seen at follow up all was well and he was able to resume his normal life pattern and activities.

Over the years, I did between four and six of these operations a year and was fortunate to have a survival rate of over sixty percent. Subsequently after the opening of the Royal Shrewsbury Hospital, patients following operation were admitted to the ITU which improved the post-operative monitoring and aided the survival of some higher risk patients.

During my first ten years or so in Shrewsbury, most of the emergency admissions resulted from appendicitis, gall bladder disease, the complications of peptic ulcer, the complications of large bowel disease and road traffic accidents. The latter could be the most distressing.

In the era prior to motor cyclists being required by law to wear crash helmets, a side room was almost invariably occupied by a young rider comatose as a result of an injury to the unprotected head: he would be under my or my fellow consultant's care. Some began to get a modicum of recovery after a few days but rarely returned to what relatives would describe as the normal pre-injury state. For the others it was a matter of managing the unconscious patient, maintaining body fluid levels, feeding through a naso-gastric tube and catheterizing the bladder to enable urinary drainage. Nursing was also stressful, the patient requiring turning every two hours and attention to skin care to prevent pressure sores and ulcers. Not only did one have the patient to care for but much time would also be spent putting relatives in the picture.

A day or so earlier a strapping young man brought to early adulthood by his parents could have set out from his home with all the joys of life to anticipate, only to be cut down by fate. Head injuries could be categorized using a formula called the Glasgow coma scale. In the worst injuries, examination could define those for whom there was no hope and who were brain-dead but being kept alive on a respirator. In this group there was the possibility of turning off the life-support machine, retrieving organs and using them for transplantation.

The relatives were first faced with coming to terms with a life-threatening condition, a very emotional experience not only for them but also the carers. Over the next few days it would become apparent that clinically there was no chance of survival whereupon the relatives would again have to be approached. The reasons for coming to this decision, always backed up by an independent clinical assessment reviewing the situation in totality, would have to be explained. One would discuss the process of turning off the life-support and then broach the possibility of retrieving organs and using them for transplantation. Realizing that there was no hope for the patient, the relatives usually gave consent in the hope that their loved one could restore some quality of life to a recipient. If consent for donation was not given, relatives often asked to be present when support was withdrawn and life expired. When consent was given the centre in Cardiff was notified, tissue-typing done and arrangements for harvesting made. This was usually done in the evening when the patient would be taken to theatre placed on the operating table and prepared as for any operation. When ready, all the checks having been done by the harvesting team, the anaesthetist turned off the life-support and on the pronouncement of death, the cadaveric retrieval was performed.

For those less severely injured, while all care was maintained, arrangements would be made for the transfer of the patient to a nursing home or occasionally a cottage hospital near the family home for ongoing long-term hospitalization. Setting this in motion and moving the patient to this facility could take weeks or even months to organize. Sadly, because of the time taken to transfer an injury, there could often be two patients in side rooms at the same time or if one had recently been moved, another victim with a similar injury would be admitted and the cycle repeated.

Eventually, with a change in the law making the use of crash helmets compulsory, the frequency of these injuries reduced to become a rarity.

In the 1970s, a considerable number of acute admissions resulted from the complications of peptic ulceration, which mainly affected the lesser curve of the stomach and the first part of the duodenum. Most patients smoked cigarettes and there was no doubt that this habit related to the development of this pathology, sufferers being mainly smokers. By this time, it was possible to pass a gastroscope through the mouth down into the stomach and both observe the pathology and obtain a biopsy for sending to the histo-pathologists for analysis. Therapeutic abilities had not yet arrived.

Two main emergency pathologies existed. These were firstly perforation of the ulcer into the peritoneal cavity and secondly erosion of an artery behind the first part of the duodenum causing haemorrhage. With the first acid leaked into the abdominal cavity giving very severe pain.

One evening, while on for emergencies, I happened to be in A & E when a man in his early thirties was wheeled into the department on a trolley. I could see he was in severe pain and went to examine him. He said that he had been in a nearby pub enjoying a drink with friends when he suddenly collapsed in pain. He was somewhat overweight and was a smoker. He gave a history of indigestion and had been having some upper abdominal discomfort relieved for a while by eating food or drinking milk. When I examined him his face was pale but his pulse rate was near normal. On examining the abdomen there was board-like rigidity. On listening to the abdomen with my stethoscope there was complete silence. These findings added up to a perforation of a peptic ulcer almost certainly in view of the history just distal to the stomach in the duodenum The diagnosis was confirmed by taking an erect X-ray of the abdomen when free gas could be seen under the diaphragm. It was then a relatively easy process to take the patient to theatre, close over the hole by plugging it and suturing in place some adjacent fatty tissue. The abdominal cavity was then washed out and cleaned before closing the abdomen and finishing the operation. It was inadvisable to do definitive surgery for the ulcer at that stage due to the possible contamination and a much increased risk of complications if attempted. The patient made an uneventful recovery and was discharged home eight days later. The need for long term curative surgery was to be assessed at the time of follow up in out-patients

Erosion of an artery, behind the duodenum, due to ulceration gave rise to severe bleeding into the lumen of the gut. Patients with this problem were usually admitted under the physicians for conservative treatment in the first instance. The blood volume would be restored and antacids and milk given through a naso-gastric tube. These measures would control some patients while others would continue to bleed whereupon a surgeon would be summoned. The patient would be taken to theatre and, on opening the abdomen, the pathology assessed. There would in these cases be no intra-peritoneal contamination. The duodenum would be opened and the bleeding point identified. Because of scar tissue associated with the ulcer, under running and securing the bleeding could be challenging and difficult. Having secured the bleeding point, definitive surgery for the ulcer could now be done.

Elective surgery for peptic ulceration was also frequent, patients requiring surgery for among other pathologies, debilitating pain and obstruction to the outlet of the stomach due to scarring producing a stenosis and near total blockage.

The treatment of peptic ulceration changed dramatically in the early 1980s following research by Doctors Warren and Mitchell in Perth, Australia. They discovered a causative organism H.Pylori. This could populate the stomach and being resistant to acid would produce ulceration. With this discovery it was not long before treatment was devised. Patients were placed on Protein Pump Inhibitors (PPIs) to reduce the acid output and antibiotics. Combined with this, there were advances in technology so that instruments could be passed through a gastroscope into the stomach and duodenum to enable cautery of bleeding points in some but not all cases. These developments radically changed the management and treatment of peptic ulceration and lead some time later to the doctors being awarded a Nobel Prize. The need for surgery for peptic ulceration became less common.

Acute gall bladder admissions resulted from two pathologies namely biliary colic and acute inflammation of the gall bladder, cholecystitis. In the former a gall stone could become impacted giving severe pain or on occasions if a stone was in the main bile duct, jaundice as well. These patients would usually undergo confirmation of the diagnosis and then on the same admission operation for removal of the gall bladder and, if necessary, exploration of the common bile duct and removal of the stone or stones.

In patients with acute cholecystitis, the gall bladder containing stones would become infected giving localized peritonitis in the right upper quadrant of the abdomen. Associated with this an inflammatory mass would form, the body walling off the infected area. This would be treated with antibiotics and intravenous fluids until the infection was under control. After a few days, once the acute episode was over the patient would be discharged home on a fat-free diet. Two to three months later by which time the inflammation would have settled the patient, more often a female, would be readmitted for removal of the gallbladder, hopefully before a further attack occurred.

Not all my time was spent in the operating theatre and doing outpatients and ward rounds. Being the newest arrival to the surgical consultant staff, it was not long before I was appointed surgical tutor. This involved organizing weekly meetings, when each firm in turn would give a presentation of their work and discuss any clinical cases of interest and of educational value to the trainees. For this there was a three line whip, everyone being expected to attend unless dealing with an emergency admission. These were useful meetings giving us all the opportunity to make suggestions in complicated diagnostic cases and to learn how others might treat a difficult problem. Often there would be alternative methods of treatment of equal merit available and this would lead to a discussion on the approach and benefit of each, again of value to all trainees. The surgical consultant staff and trainee registrars all hailed from different alma maters, where often approaches and treatments might have varied. The case histories and management of patients that had died were also presented to determine if there was anything to learn and perhaps improve the treatment of any future patient with a similar condition.

Other duties of the surgical tutor included attending regional meetings in Birmingham twice a year when tutors from all hospitals throughout the region would be present. Inspections of all hospitals throughout the region would take place on a regular basis. A representative from the Royal College of Surgeons and two surgical tutors from within the region would visit a hospital and meet the local consultants and junior staff to ensure the facilities and teaching arrangements were up to scratch,. Everything from looking at theatre and ward facilities to looking at accommodation provided for junior staff and inspecting the medical institute and its postgraduate program would be examined. Seeing and cross examining junior staff was an important part of these visits. Following this, the visiting party would discuss the findings of the meeting and then meet up again with the local surgical tutor and present our views as to where if necessary improvements might be made. The college representative would, within a few days, forward a copy of his report and recommendations to his fellow examiners for agreement or revision before sending on to the College. Most inspections were fine but occasionally short-comings would be unearthed in which case a further inspection would be recommended to take place the following rather than in the normal three year cycle.

Twice a year, surgical tutors from all over the country would attend meetings at the Royal College in Lincoln's Inn Fields when we would be updated on College policy and training developments. The meetings would be chaired by the President and advice given on programs for the trainee staff. On one occasion, there was guidance on what action to take if we found a trainee inadequate and not suitable for a surgical career. This could precipitate a difficult and emotional state when we not only had to consider the trainee but more importantly any outcome for a patient who might be treated by a substandard junior. I had an example of this problem on my unit, when I became aware it was necessary to monitor the incumbent more closely and only allow him to proceed with the most simple and straightforward cases. I was hoping on this regime there could be an improvement, in which case his training could go forward. If not, the situation would have to be put in the hands of the Regional Advisor, the College representative in the West Midlands. The problem in my experience appeared to be blindness of the individual to his own shortcomings and lack of appreciation of his inadequacies. After a number of months and hoping for improvement, I had to refer him to the Regional Advisor who decided to give him one more chance. The individual in my case went on to another centre, where the same problem having arisen, he was suspended and advised he was not suitable for surgical training. This was a sad experience but quite rightly the well-being of patients had to be put first.

After the meeting in the morning, there was a luncheon at the College when there was the opportunity to catch up with one's contempories and one's previous trainers some of whom were now on the College Council. Communication was all important to ensure you kept up to date with advances in your field and didn't become fossilized buried away in the country.

Other means of keeping up to date were by joining the Association of Surgeons of GB and Ireland and in my own speciality the Vascular Surgical Society. Both these august bodies had annual meetings lasting up to three days when papers outlining advances in management and techniques and results of new treatments for patients were presented. The research projects produced by academic units would be reported and again give help and advance our knowledge in the management of various conditions. Inevitably in one's own practice one could come up against a problem and it was an advantage to be able to discuss this with one's peers both giving and receiving advice.

From all this, it will be apparent that nothing was stationary and all the time progress was being made not only in surgical technique but in instrumentation and technology. There were improvements in anaesthesia, in the ability to suppress tissue rejection, allowing expansion in organ transplantation, and in post-operative care, the ill being treated in intensive care units. As with so much in everyday life great advances were being made, in this case for the benefit of the patient helping to improve longevity and the quality of living.

Chapter 11

Changes in Shropshire

The workload escalated as I found my feet and my predecessor wound down. I soon found that I was being asked to do a lot of domiciliary visits. Those nearby tended to be single while with those further afield, such as in Oswestry or Radnorshire, the GPs would usually ask me to see an emergency or urgent patient and, at the same time, get me to see other patients who might otherwise have had a lengthy journey to hospital outpatients. Doing this, the GPs were no longer anonymous and it enabled me to build a rapport with them particularly useful if, when patients were discharged from hospital after treatment, they went to a local cottage hospital under the GP's care. These facilities enabled the elderly to recover and get mobile in a local environment where they could be visited more easily by their family and friends, often of a similar age. The nurses, fully qualified, and their assistants were usually part of the local community and a physiotherapist was often available. These facilities allowed those, especially the elderly and those living alone, requiring a longer period to recuperate, the opportunity for a more leisurely recovery. With the patients' needs considered and fulfilled, it meant for the hospital that a bed was no longer blocked and the considerably more costly facilities could be used for the benefit of the next patient on the waiting list.

The variety of conditions requiring surgery was large and I was ever grateful for the breadth of my training, which is no longer possible today, when entrance into a speciality is early and operating experience even at the end of training by comparison limited. Knowing how long on average it took to do a particular operation and often knowing the patient, it behove me to select the patients from the waiting list for admission. I also understood the degree of urgency for treatment. Once established as a surgical consultant, GPs would contact me or my secretary to expedite a patient on the waiting list for admission, if there had been deterioration in the meantime in their condition. In order to utilize all the time available, I would organize a list on the basis that all would proceed smoothly. This on occasions could cause problems when an operation proved not to be routine and far more complicated than anticipated. As in the main most operating lists progressed smoothly and as expected, the odd overruns were tolerated by the theatre staff, it rarely being necessary to cancel patients prepared for operation.

Routine lists in theatre could on occasion allow chatter. One May, I was operating on a youth to remove a diseased testicle, performing an orchidectomy, to give it its surgical name. A lovely Welsh nurse, Blodwen, in her twenties, my scrub nurse for this case, reprimanded me in a broad Welsh lilt for not bringing her a present as it was her birthday that day. I told her that it was mine the day before and that I hadn't heard from her. At this point the operation came to its climax and I was able to reassure her that I had something for her, presenting her with the orchid that I had at that moment removed. This caused audible amusement and staff came from outside the theatre to join in the mirth.

In the late 1970s, there was a change in the surgical staff when two senior members retired. An urologist was replaced as was a general surgeon, the latter appointee having done his postgraduate training in Birmingham, where in addition to general surgery he also had gained vascular experience.

I continued to get an increasing number of patients with peripheral vascular disease. Operations on patients with this pathology tended to take longer than those with general surgical pathology and those undergoing surgery to save a limb could be demanding. Operations for patients with hardening of the arteries carried a better outlook because usually the vessels below the knee into the lower leg were patent when it was a matter of bypassing the proximal obstruction to re-establish flow. This could be done by using the patient's own tissue in the form of a vein harvested from the leg or by inserting a Dacron tube graft. A third alternative could be reaming out the diseased core (an endarterectomy) leaving the outer wall of the vessel to form a patent channel.

In another group of patients, vascular occlusion would be associated with long standing Diabetes. In this group, the vessels below the knee were often of poor quality making the surgery much more challenging and the outcome less favourable. Although pulses may have been palpable and the colour of the foot good at the end of the operation, early occlusion, once the anticoagulation needed for the procedure had worn off, could occur. This would necessitate within a few hours a return of the patient to theatre in an effort to save a limb. Not all salvage attempts were successful, sadly necessitating amputation as a third procedure. Later on surgeons began to place grafts from the knee down to the foot. These procedures were particularly demanding when trying to join on to much smaller vessels in the feet and seemed of limited success.

In the 70s and early 80s, before the advances in the treatment of peptic ulceration, a lot of operations were done to treat patients electively with severe symptoms. These in the main consisted of obliterating the nerve supply to the stomach to diminish acid output and performing a drainage procedure to allow the progress of the gastric contents on into the gut, the valve at the stomach outlet no longer being functional following nerve section.

One day, the gastroenterologist came to see me as a patient. He was suffering from severe heartburn, acid from his stomach regurgitating into his lower gullet and causing his symptoms. After appropriate investigations, confirming the pathology, he proved a suitable candidate for the operation I had learned and performed and written about at St. Thomas'. His wife, who was the sister in charge of the theatres, had obviously vetted me and considered me suitably qualified and skilled to operate on her husband.

On the day of surgery again by request, I had Kay Hurdman, the London trained anaesthetist, to put him to sleep. While preparing to operate, I reminded myself that he had been referred because of the success of operations that I had done previously on other patients with similar pathology and that it was important to do the procedure in the same way, with as far as was possible no variations. I was fully aware that I was operating on a colleague and friend as I incised through the skin but once into the abdominal cavity my mind cleared of these thoughts and I concentrated on the problem in hand. I found that the hiatus or channel by which the gullet passed through the diaphragm, from the chest to the abdominal cavity before entering the stomach, was enlarged. This easily allowed the lower gullet and adjacent stomach to ride up into the chest where the change in pressure held the "valve" open allowing acid to reflux into the gullet. As in previous similar operations, I drew the stomach down and tacked, with stitches, the anterior wall of the stomach to the periphery of the diaphragm, thus preventing any further migration into the chest. The operation was completed and after recording the operative procedure in the notes, I rang his wife to report all was well. I remained in contact with this patient for many years and I was pleased to observe that he remained symptom free of heartburn.

During one Christmas holiday being on call, I was phoned early on Boxing Day to say that a nurse, who had been on night duty, was in the A and E with abdominal pain, having been involved in a road traffic accident while on the way home. Her car had skidded on ice while going over a bridge and she had had a bang on her left side over her lower ribs. She appeared to be bleeding internally and the site of the injury suggested she had damaged her spleen. Further observation confirmed continuing blood loss and it was necessary to take her to theatre where, on opening the abdomen, I confirmed a ruptured spleen necessitating splenic removal. Not a good Christmas present. Thankfully she went on to make a good recovery.

Not all medical emergencies occurred in hospital. One evening, while at home, my wife developed severe right-sided abdominal pain radiating from the right loin down to the groin. She had developed renal colic and I had to call our GP who visited and prescribed analgesics to settle the acute pain. Subsequent investigation revealed an abnormally developed kidney not fit for purpose other than giving trouble. An ex-colleague at St. Thomas', then the senior urologist in Wolverhampton, kindly saw her and removed the offending organ. This gave me some close at hand insight into a patient having an operation and, equally important, the process of recovery and rehabilitation after surgery. If I hadn't comprehended fully before, I realized subsequently that when muscle was cut, to allow the surgeon access, it would take six to eight weeks for the muscles to heal fully and the wound to become pain free. Only then would the patient get back to feeling normal and capable of returning to full activities.

In the mid 1970s our daughter developed a chest problem sufficiently worrying and urgent to require the help of a colleague, a Chest Physician, whom we visited on a Saturday afternoon. He diagnosed asthma. When he was taking a history, I remembered that my mother had always been chesty. On my mother's side there was a history of her grandmother having twelve children, only two of whom survived to adulthood. Almost certainly the reason for this would have been asthma. During the following years, we had many narrow escapes with our daughter, the attacks coming on suddenly out of the blue and necessitating dropping everything and jumping in the car and rushing her to hospital, on occasions straight to the ITU. Jack, the Paediatrician, was marvellous. We would ring him and after asking a few questions he would say take her to the Paediatric ward or the ITU. While speeding our daughter into hospital, he would notify the department that she was on the way in. On arrival at the hospital some five minutes later she would be struggling for breath and blue around the lips. She would be placed immediately in an oxygen tent to buy time and before the necessary medication prescribed took effect. Because of the sudden onset and severity of these attacks we realized that we would have to live within a short distance of the hospital to gain rapid access. It was then that we understood how fortunate we were to have moved to Shrewsbury where traffic congestion was unknown. These bouts proved physically and emotionally draining, disturbing and frightening for all and it would be a day or two before the effects of the shock would wear off. But for having a mother who had trained as a nurse and a father in the medical profession it is probable that she would not have survived to adulthood when, although still present, the severity of the attacks diminished.

In 1980, we had a third child, another daughter, Charlotte, born on a Sunday. I was at the birth and all present at the delivery were amazed as she was looking round at everyone before being fully delivered, as if to say I'm here what have you got to offer. She proved a very active child with abundant energy. One day before she was two I arrived home to learn that she had been having paroxysms of pain when she would scream and draw her knees up towards her chest. I watched the next attack and observed these features. When the pain subsided after a few minutes she was pale and limp. With my experience at Great Ormond Street, I recognized these symptoms as being indicative possibly of an intussusception, a condition where the terminal small bowel can telescope into the large bowel causing a blockage. On examining her there appeared to be some emptiness in the right lower tummy in the appendix region, a sign which if correct would confirm the diagnosis. I rang the GP who sounded a little sceptical but after discussion suggested I take the next step and contact a radiologist. I rang Michael Dean who happened to be one of my daughter's godparents and was even more sceptical but said he would come in and to put my mind at rest do a barium enema. I delivered her to him and returned home to await his call. After more than hour, Michael phoned, somewhat agitated, to say that indeed I had been correct and that she did have an intussusception which, with some considerable difficulty, he had managed to reduce. He hoped all would remain satisfactory but confirmed that there was a five to ten percent chance of recurrence. But for my being in a position to make the diagnosis, treatment would have been delayed for a few hours by which time open operation would have been mandatory. In all probability my daughter would have required a partial resection of her gut and grown up with a scar on her abdomen. Fortunately she was one of the ninety percent who went on to make an uninterrupted recovery.

The lesson that I learned from these experiences was that it was possible for me to turn over from family to doctor mode and take a detached view of the circumstances that I faced.

With the arrival of the third child we had outgrown our house on the Mount and needed to move. We were fortunate to find a property in the country at Cruckton on the Montgomery road, the same road out from Shrewsbury as the hospital. If anything this was even nearer the hospital than the first home and in view of the circumstances with our oldest daughter an important factor.

By the time of our removal in 1981, my parents were in their late seventies and frail. My father was suffering from Parkinson's Disease and was physically infirm. Mentally he had periods of rambling and being incoherent while at others communication would be normal. It transpired, that while rambling, he could understand and remember all that was said to him which proved hurtful, as some visitors failed to be aware that he could comprehend. He was still interested in all that I was doing and was full of enquiries and capable of sound advice. After the onset of his disabilities, on his last visit to Shropshire in the late seventies, we took him to lunch at Lake Vyrnwy, in Mid Wales. When he discovered that the waitress spoke Welsh, his face lit up and in what proved a very happy and enjoyable outing, he used his native tongue in conversation for the last time.

In 1982, over the August Bank holiday, we were due to visit my parents in Haslemere, my father then being in a nursing home. My mother phoned the night before we went and informed us that my father had died that evening. It proved a sad journey down. My father had been an inspiration to me and the light that led me to a surgical career. He had nurtured and developed my early enthusiasm and seeing that I took to the discipline like a duck to water had, from an early age, enabled me to participate and acquire some basic operating skills. The weekend was not all gloom as we were able to tell my mother that we were expecting a fourth child.

In 1979, the Royal Salop Infirmary shut, and all the facilities, except those at The Eye Nose and Throat Hospital, moved to the site on Mytton Oak road. This included the Outpatient and A&E departments which were now just a short walk across the road from Copthorne Hospital and a Day Case centre with an operating theatre. Additional facilities, a helicopter landing pad and a conference centre, were added. These changes meant that nearly all the staff were on one site and could meet at lunch in the Medical Institute, giving the opportunity to discuss problems as they arose.

One day, the microbiologist discussed the place of prophylactic antibiotics in patients undergoing large bowel resections for both inflammatory and malignant conditions for which, at that time, there was not a uniform policy. A trial was set up and patients randomly allocated to undergo different regimes. In all, observations were kept on seventy patients with an equal sex distribution. We found that a single dose of one antibiotic, rather than a single dose of two antibiotics, was equally effective in preventing post operative wound infections, except when the operation was for inflammatory bowel disease when two antibiotics and a longer course of treatment was indicated. Once more, we found we could hold one antibiotic back in reserve and help prevent the development of antibiotic resistant organisms. Not having participated in a research project since leaving St. Thomas', it was stimulating to be able to answer a question and set up a sound treatment protocol for patients undergoing these procedures.

During the 1980s, with the expansion and development of Telford giving an increase in population, a decision was made to expand the hospital facilities in the county. After a considerable amount of debate a decision was finally reached. With the Royal Shrewsbury Hospital only recently constructed and opened, it was felt that this could not be sacrificed. One option had been to build a major hospital between Shrewsbury and Telford serving both towns and still fulfilling access to the hospital from within the boundaries of the Area Health Authority. It was decided eventually to build on a new site in Telford which with the benefit of hindsight and experience was probably the wrong decision, as now some twenty-five years later services are split between the two sites, some facilities being in one and not in the other, rather than all under one roof.

To cope with the increasing workload and pending the opening of the new hospital further appointments were made to the staff. There were now six surgeons. Throughout this decade, with these new appointments the work pattern changed. Whereas when I arrived there was a lot of urology, this diminished with the appointment of an urologist. One of the last transurethral resections of prostate that I did nearly ended in disaster. A man in his late seventies was admitted and underwent what appeared to be a routine operation. After the operation he haemorrhaged, the blood passing down his catheter to the exterior. This continued and I was faced with a life-threatening situation. He was taken back to theatre and required an open operation to stem the loss. The situation was tense and the anaesthetist was giving blood transfusions to maintain the blood pressure. To gain control, an incision was made in the lower abdomen. The prostatic capsule was opened and the raw area packed. Because of his condition he was taken to ITU for the night and fully resuscitated. His treatment was then completed and he went on to make a full recovery. It transpired that because of a heart problem he had been taking aspirin on a regular basis to thin his blood. When asked pre-operatively whether he was taking any drugs he had answered no, not considering aspirin which he could buy over the counter a drug. The anaesthetist had had the same response. After this hair-raising experience both for the patient and the surgeon, it became my policy thereafter when taking a patient's history always to enquire whether they were on aspirin.

As the decade advanced and the opening of the Princess Royal Hospital approached services were rationalized. The last appointee as a surgical consultant was to move to Telford so that we returned to five. With the opening of the hospital, the surgical services in Shrewsbury consolidated in the Royal Shrewsbury those that had been in Copthorne transferring over the road.

In 1983 when well settled into our new house, we had a further child, Sarah, a daughter, who was destined to become a doctor being the fourth generation to enter the medical profession spanning a period of more than one hundred years. From 1982 to 1985 in turn my father, my mother, my father-in-law and my mother-in-law all died so we had a sequence of years with grief.

Surgery despite the increases in staff continued at a pace there being a never ending stream of referrals both within and outside the health service. The hours worked remained much as they had been in London, the only difference to the day being the short drive to and from work. In the mornings, in their early years, I was able to take the children to school but by the time I returned home they were usually well into bed and asleep. At weekends I usually returned to the hospital to do a ward round and check on all those who had had an operation that week. One Saturday, while in Copthorne hospital before we moved over the road there was pandemonium when one of the hutted medical wards caught fire and became a shell in little over an hour. Happily all the patients were evacuated to the exterior, it being a sunny early summer's day, and there no casualties. The fire service arrived promptly and were able to contain the fire to one ward there having been a considerable risk of it spreading to the whole hospital. The building was reconstructed and modernized to form an endoscopy unit allowing both out and in patient investigation of the gut from mouth to anus.

During these years, David and I were fortunate to have Australasian registrars coming to this country to get "cutting experience". The quality of these was excellent, they coming to this country as the equivalent of Senior Registrars before returning home to get Consultant appointments. One of these while here gave rise to an amusing incident. While still a student, he had been unfortunate, while riding his motor cycle one New Year's Eve in Sydney, to have an accident necessitating a below knee amputation. At weekends when not on duty, and having purchased a bike, he would put his motor bike on a train and visit differing areas of England. One weekend in London on a wet road he skidded at traffic lights when coming to a halt. He fell to the ground and onlookers rushed to help. He was unhurt the only casualty being a lady who on seeing him on one side of the bike and his leg on the other side promptly fainted.

During these years, my radiologist friend Michael Dean continued to advance his skills, not only obtaining excellent films of the vascular tree from the abdomen to the feet but starting initially with patients unfit for a lengthy operation to carry out interventional procedures. In the course of time he became adept at inserting percutaneous stents and or balloons to treat narrowed areas or short blockages of diseased arteries and so restore arterial blood flow to the lower limb. He progressed slowly but getting good results moved to treating patients otherwise fit for open operation. Apart from operating on aneurysms and diseased major abdominal vessels, percutaneous procedures became the treatment of choice, patients' recovery times being measured in hours rather than days.

Treatments continued to evolve and the need for surgery for peptic ulceration and now elements of peripheral vascular disease were fast disappearing from the surgical repertoire. I was beginning to wonder what I would be doing for the next ten years of my medical career.

One day towards the end of 1989 a young woman aged twenty came to see me accompanied by her mother. Her history and findings on examination indicated that she had a form of gall bladder disease giving rise to recurrent and severe bouts of biliary pain. I explained that she needed her gall bladder removed to cure her symptoms. The mother then said she had heard of a procedure which did not involve a large cut in the upper abdomen. I had recently returned from a national surgical meeting and had heard nothing regarding this while in attendance. The woman's symptoms were quite severe interfering with her normal life and I advised early operation which was consented. The gall bladder was removed through a moderate incision but nevertheless left an obvious scar in the right upper abdomen. The reason for discussing this case will be revealed shortly. The scenario haunts me to this day.

In the Spring of 1990, I received a phone call from David Rosin, an old colleague in London, inviting me to a meeting in St Mary's Hospital in May to learn about a new form of surgery. Being surgical tutor, I readily accepted and duly attended the meeting which was to change my career and become a milestone in the treatment of patients.

Chapter 12

The Road to Damascus

The meeting chaired by David Rosin began on a Monday morning in May 1990 at St. Mary's Hospital, about one hundred surgeons being in attendance. An American surgeon, Jo Petelin, from Kansas City was introduced to us. Almost immediately a patient who had had her gall bladder removed three days earlier, on the Friday, was introduced to us. A buzz went round the lecture theatre as she walked in, her bearing being completely normal. All present in the room were accustomed to a patient after a gall bladder removal just beginning to mobilize at this stage with a view to returning home eight to ten days after operation. She was questioned and said she was up and returned home on the Saturday, one day after surgery, and that once the anaesthetic had worn off, apart from a slightly distended tummy, had suffered no ill effects. We were all amazed. As soon as the patient retired, another was introduced, again looking well and sprightly, and repeated the same story saying that she also had had her gall bladder removed on Friday and had gone home the following day.

Jo began his talk by saying that a Gynaecologist, Professor Mouret, in Lyon in 1987 had a patient in her early fifties who required an operative procedure for a gynae problem but was also suffering from attacks of biliary pain due to gall stones. She asked if the two operations could be done at the same time. Gynaecologists had been looking into the abdomen with telescopes for some time and over the years cameras were mounted on these and instruments to perform some minor operative manoeuvres were developed. The surgeon began his operation by inserting four litres of air into the abdominal cavity to create a pneumo-peritoneum and room to operate. Once inflated, three portals fitted with valves to prevent the extrusion of gas were inserted. Through one, the telescope with a camera mounted and connected to a video screen was placed while through the others instruments required to perform the procedure were passed. Having done the gynae task, the surgeon looked into the upper abdomen and found that the gall bladder was clearly visible. After inserting two more portals through five mm. incisions he found he was able to clip and divide the vessels and duct to the gall bladder and remove the gall bladder from the liver bed without undue difficulty, before delivering it to the exterior. The patient made an uneventful recovery whereupon excitedly he phoned his friend, Jacques Perissat, Professor of Surgery in Bordeaux, who took on and further developed and advanced this technique. In 1988, Perissat spoke at the SAGES (Society of American Gastrointestinal and Endoscopic Surgeons) meeting in the USA presenting this laparoscopic or keyhole operation to the Americans. Following this it was taken up by, in particular, Dr. Eddie Reddick of Nashville, Tennessee and among others Jo. By the time he came to this meeting Jo had done in excess of one hundred keyhole removals of the gall bladder. After detailing how this was done and showing some video clips of his operations, we were invited to go to another room where some simulators were set up. These consisted of boxes with an opaque foam top. Inside there were objects to move and tubing to clip and divide. A portal with camera mounted and connected to a screen was first inserted and two further portals pushed into the box to pass instruments to carry out the exercise. When it was my turn a Rep from one of the main companies associated with instrument development and production happened to be watching. It was not long before he said that I seemed to have a natural ability to perform the tasks and that I should take it up. After lunch David Rosin invited me to go to the operating theatre and watch Jo perform. This was fortunate as the rest of those attending the meeting returned to the lecture theatre to view the operations through a video link. By this time I realized that I was travelling down the road to Damascus and that I was viewing the future of surgery. While seeing how it would benefit patients needing gall bladder removal, not having as yet done an operation, I was not able to appreciate how far these techniques would replace what had been conventional surgery. Viewing Jo operating close at hand, I was able to see what was required regarding equipment and how it fitted into the operating theatre and what instruments he was using and why. The other amazing factor was the magnification supplied by the telescope and camera up to more than ten times that seen with the naked eye. There were two major surgical companies Tyco Healthcare and Ethicon-Endo Surgery, who were leaders in this field at that time, although there were others not far behind.

I discussed, with one, the process and costs of setting up and equipping to do these operations. It transpired that to get tooled up and trained ready for take off would require one hundred thousand pounds.

I left the meeting exhilarated and excited and determined to get off the ground. I contemplated even trying to put up funding myself, an idea immediately and quite rightly vetoed by my wife who said family and education must come first. I talked around the hospital to see if there was any money in the equipment budget or in the region to finance this project. After a while I was recommended to try the League of Friends of the hospital who could help out in worthy projects when funds were available.

Around this time and unbeknown to me a Shropshire farmer while holidaying in Sydney collapsed and died from a heart attack. I had inherited his wife, who had been treated at onset by my predecessor, for cancer of the breast and, with intermittent treatment, had kept her in reasonable health for some fifteen years before she died. He had also been a patient of mine requiring repair of a groin hernia. In his Will he left two hundred and fifty thousand pounds to his family and the remainder, some four hundred thousand, to the League of Friends. At first, the family thought of disputing this but soon seeing that the taxman would take a large amount and that they would each benefit very little, decided to let things stand and uphold his wishes.

At about the same time, the number of beds that were available to admit patients for operation was cut which once more was going to place an enormous and almost impossible burden on the waiting list. I went to the Chair of the League, June Whitaker, and told her what I had seen and what a benefit it would be to patients to have the means to perform these operations. I also acquainted her with the bed situation. Following this meeting I was asked to attend the next committee meeting and put my proposals to that body. I duly explained to the assembled gathering what I had seen and what I proposed.

I was asked a number of questions, the most searching being as to whether I thought that once supplied with the equipment I would be able to use it. I replied, with some temerity, that having been on a simulator and the favourable comment made at the time, I thought it would be manageable. Furthermore, if the committee granted the funds I would be taken to Kansas City with others to observe and watch Jo in action and pick his brain over the course of a week. After a few moments, the Chair moved that I be supported in this project and there was unanimous agreement.

I was elated; as not only would patients benefit but I also would have a new challenge and be in at the dawn of what I was sure would be the way forward in the field of general surgery.

In October, four of us and a Rep from Tyco Healthcare (now Covidean), our organizer or tour leader, assembled in London and flew to Minnesota. Once there we transferred to an internal flight to Kansas City. Of the others on the trip, I knew George who had trained at St. Mary's Hospital and was a surgeon at Worthing, where one of my brothers was a Physician. This was my first trip to the States and as a visitor, I liked what I saw. We were billeted in a tall hotel, our rooms being on the tenth floor of a twenty-storey building.

We had a day to acclimatize before meeting up with Jo and becoming his shadow for the rest of our stay. During the week, we watched him operate and remove gall bladders from some ten patients, all the time remembering that we ourselves would be equipped and operating within a few weeks. I was conscious that when that day came, other than the Rep who had watched some thirty or so operations during a number of visits, there would be no long stop to refer to in the event of difficulty. We needed to watch each operation from A-Z. In particular we studied how he inserted a needle into the peritoneal cavity before inflating it with air to create a working space or pneumo-peritoneum. We then watched where he put the access portals, usually three, and discussed the reasoning for their positioning. Once inside and set up, we looked to see how he manipulated and displayed the anatomy marvelling once more at the vision obtained with the magnification. Again I studied the manoeuvres to mobilize and free the gall bladder for removal to the exterior. To ensure it was removed whole and to prevent what could be hundreds of small stones spilling into the abdominal cavity, the gall bladder would be placed in a bag to deliver it through the abdominal wall to the exterior. Once out, the abdomen would be decompressed releasing the air to the exterior. It was then a short process to close the puncture holes none more than one cm. in length and complete the operation. After four days of watching and discussion with Jo and between ourselves it was time to return to the UK, armed with memories and videos to prepare for the next step.

There was an amusing incident on the way back. George told me that on the next day in London he had to give his daughter away and make a speech at her wedding. He was worried that he wouldn't sleep during the ten hour flight. It so happened that I was carrying a sleeping pill. The plane was three quarters empty so we could lie down across a row of seats. George took the pill and shortly afterwards food arrived. George ate his main course and then slumped asleep. The air hostess arrived concerned in case she had a medical emergency on her hands. I was able to reassure her as to what had happened so we laid him across the seats to continue his sleep and she cleared his tray away. Some hours later he came to. In a sleepy voice, the first thing he said was: "Where's my pudding?". On Saturday morning George left the plane refreshed and ready for the afternoon nuptials.

On return to Shrewsbury, there was a short time to get set up before the equipment arrived. I got a simulator box so that I could practise under direct vision the manipulation of objects using instruments which were forty five cms. long. In the evenings before retiring I repeatedly played the videos so that I could almost do the operation in my mind. Meanwhile, while waiting, I went to the Gynae. theatre and under guidance from a gynaecologist placed needles in the abdomen and created a pneumo-peritoneum. At the beginning of December, we were ready to go. I selected two female patients for operation and interviewed them. I had to point out that I thought they were suitable candidates for this new procedure which, if successful, would enable them to go home the day after surgery and be ready to get back to full activities in two weeks rather than two months. I told them I had not done the operation in this way before, but had been on a training course and had received good instruction. I said that I was reasonably certain I would be successful but if a problem arose I would convert to an open procedure which would, in fact, be back to the old operation.

On Tuesday 18th December, I arrived in theatre and the first patient in her late thirties, the older of the two, was anaesthetised and put on the table. Ringing in my ears was the question from the League of Friends: "if we get the equipment, do you think you will be able to do the operation?" The Rep was in attendance and able to give encouragement. I started by inserting the needle and creating a pneumo-peritoneum. The portal to allow the telescope was inserted with care as if I pushed through the abdominal wall too hard, it could suddenly enter the abdominal cavity and then with minimal resistance advance rapidly and cause severe damage to the abdominal contents. All went well with this and having removed the plunger (trocar), the telescope was inserted. After an initial look around to ensure all was well I turned the lens towards the abdominal wall and chose the sites of entry for the instrument portals. I was now operating in space looking at a two-dimensional screen. The first time grabbers and dissectors were entered I was floundering finding some difficulty with orientation and placing them in the precise position that I required. After a few minutes I achieved what I had set out to do, this manoeuvre subsequently after I had done a few operations taking a matter of seconds, the instruments being directed immediately to the required position. Once there, traction was applied to the gall bladder to give exposure to the necessary anatomy. As each step was completed I was encouraged by those around me. The vessels and gall bladder duct were defined, clipped and divided. Using an ultrasonic machine the gall bladder was separated from the liver bed and freed. I then spent time attempting to place it in a bag for removal to the exterior. Having inserted a drain, the bag was extricated and the operation completed, to applause from onlookers. From start to finish had taken just under three hours. I was grateful for the training I had had and the practice I had put in to arrive at this point. The patient soon came round and all being well she was returned to the ward.

After a break, I spoke to the second patient, who was in her late twenties, before she was put to sleep, telling her the first patient had had her operation and that all was fine. I repeated the same procedure that I had done in the morning and on this occasion, with the experience of the first case behind me, found that I was better orientated and the instruments went more quickly to the desired place. After two and a half hours the operation was completed successfully and the patient recovered, before being returned to the ward. I had broken the ice and after checking the patients to confirm all was well, I returned home. Later that evening, when at home, I rang the Chair of the League of Friends to give her the news that we were under way and that the first two operations had been completed without a problem. I felt jaded but exhilarated that some nine months after setting out on this road we had finally got equipped and had started. This was indeed the beginning of a new era.

The next day when on my ward round I saw both patients. Both were up and feeling fine. I discharged them home with strict instructions to do very little for the next two weeks until I had seen them in the first outpatients clinic in the New Year. At their consultations, both were well and had had no problems. The portal wounds were well healed and would leave scars, within a month or two, which would be barely visible to the naked eye. What a change from the previous era and my mind could not help going back to the twenty-year old I had operated on one year previously who was scarred for life. On asking what they had done in the meantime the older patient had taken it easily but had felt fine. The younger one had gone to an office party three days after her surgery and, also feeling fine, had danced the evening through. I had learned another lesson. The body will regulate what you can do and, within reason, activity should be regulated by how you feel. With keyhole operations there were no muscles incisions to knit together, or stitches that could tear out. What a benefit for the patients and as an offshoot what a benefit for the hospital waiting list. A bed that previously would have been blocked for at least a week could now be used for three patients in that time.

Chapter 13

Keyhole Advancing

1991 developed into a year of consolidation when in addition to performing my general surgical commitments, including peripheral vascular surgery, the number of keyhole procedures done steadily increased. I soon added another feature, cannulating the bile duct prior to division and injecting radio opaque dye. This enabled X-rays of the main bile duct to be taken to determine whether or not stones had passed from the gall bladder and into the duct to threaten a blockage and jaundice.

Following my first trip to the US to shadow Jo Petelin I returned for a holiday in the summer with my wife and two youngest daughters. On this visit we flew to San Francisco and after a few days exploring and visiting the sites, including Golden Gate Bridge and a boat trip round Alcatraz, we journeyed by car to Yosemite National Park. There we marvelled at the vast granite cliffs and the giant Sequoias the like of which we had never previously seen. From there we drove to Monterey and Carmel before driving down the West Coast of America along route 101. In Los Angeles a visit to Disneyland was mandatory and caused much excitement. From there we flew to Las Vegas moving on immediately to the Grand Canyon which we saw in all its glory taking a scenic tourist flight. We rounded our trip off by flying to San Diego and meeting up with a semi-retired American Vascular Surgeon, John Bergan, who was attached to a hospital and lived in La Jolla , a suburb of the city. I had met John on a number of occasions at Vascular Surgical meetings in the UK and also at meetings in ski resorts in Austria, where business was combined with pleasure. John was a member of the San Diego Yacht Club, host to the America's Cup on a number of occasions, and invited us for a day's sailing round San Diego Bay, home to the American Pacific Fleet. We arrived at his mooring in mid-morning, Mary having told me to be prepared to help with the sails. After an aperitif or two John said "Let's go for a sail" whereupon he pushed a few buttons and up went the sails, my only contribution being to help cast off the mooring. After sailing round the bay, inspecting those elements of the fleet in harbour, we retired to the Yacht club for a most excellent meal before departing, our daughters having behaved in an exemplary fashion and impressing our host with their social graces and interest throughout the day. From San Diego, after a most enjoyable trip, we flew back to the UK.

During the summer of 1991, I took a day off to play with the St. Thomas' Hospital golf society. In the morning round, I played with Malcolm Gough who at one time, while I was a junior trainee, had been Reader on the Surgical Unit at St. Thomas'. He was now the Paediatric Surgical consultant at the Radcliffe Infirmary, Oxford. He knew me as a Vascular and General surgeon. I told him I had taken up Laparoscopic surgery and of the benefits it gave the patients with regard to recovery time and shortened hospital stay. After a few minutes he turned to me and said "you are wasting your time. This will be a five day wonder before it folds". Many years later, after we had both retired, I met him again at a St. Thomas' evening at the Old Vic, when a Thomas' group of national fame, Instant Sunshine, were giving a concert in aid of a charity. As I approached him, while he was talking to my old and revered boss Frank Cockett, he threw up his hands and apologized saying he had been wrong in his prediction. We both remembered that day on the golf course some ten plus years before.

As the year progressed and my skills advanced I was visited from time to time by the Tyco Rep. About the middle of the year he began to bring other surgeons from around the Region to witness the procedure. One day, I received a call from David Rosin to ask if I would join a small group who were to go to the Tyco Headquarters in America and then on to a facility where we could explore other laparoscopic techniques. There were six of us on this trip including David Rosin, Lord McColl, the professor of surgery at Guys Hospital and David Dunn a surgeon at Cambridge. On this occasion, we travelled in comfort and on arrival in the States were driven to

Norwalk to the north of New York where we stayed before going to the Tyco Empire the next morning. There, we had a conducted tour round observing instruments being manufactured and then a quick look into the R and D section, the home of things to come.

Later in the afternoon we travelled down to near Washington. That evening we were taken out to dinner travelling in a Stretch Limo, a first time experience for most of us, to the restaurant. Next day we went to a facility not available in the UK, where we could operate on anaesthetised pigs trying out new operations and equipment. The Anti-Vivisectionists had blocked this facility in the UK, the opportunity to practice on animals rather than patients only being available in Europe, in Eire, France and Germany. This was surely a safer way forward for most aspirants of laparoscopic surgery, than beginning under stress in the operating theatre. This was a major advance as, unlike when operating on cadaveric specimens, any error would cause a leak or red out. We tried new instruments, including laparoscopic staplers, and practiced suturing and tying knots. By now, we in the group believed that keyhole removal of the gall bladder was only the beginning of the story and that learning on an animal model was vital before subjecting our patients to other laparoscopic techniques. We were learning a new approach to operations that previously we had had to perform open. After our day there, we were due to go to Washington to catch a plane back to the UK. It was then that our guide learned that the flight had been cancelled. For a time it appeared that we would have to get the Concorde flight, where seats were available, out of Washington that evening. Sadly, this was the nearest we came to travelling on this plane as it was discovered that there was a flight from Washington to New York which we took and connected with the red eye flight to London.

On this trip, it became apparent that a new society would need to be formed to promote laparoscopic surgery in the UK. Thus the Society of Minimally Invasive General Surgeons (SMIGS) was born. SMIGS had its first meeting early in 1992, by which time we appreciated that there was another group in the North of England, the British Surgical Stapling Group (BSSG) which had evolved into Keyhole surgery. Considering it was a new society, the first meeting was well attended and it was at this that forthcoming third World Congress of Endoscopic Surgery was announced. This was to be led by the European Association of Endoscopic Surgeons and SAGES (Society of American Gastrointestinal and Endoscopic Surgeons) and was to take place in Bordeaux under the Presidency of Professor Jacques Perissat.

The meeting convened in June. Never before or since in my career did I witness at a surgical meeting, and I attended two to four a year, such scenes of enthusiasm and excitement. There was a continuous buzz as delegates marvelled at what was opening up before their very eyes. Surgeons of all ages were literally running, like excited school boys, from one lecture venue to the next so as not to miss out. Conversation was very animated as we now began to understand fully the extent to which we as surgeons were going to be able to affect patients' lives. An afternoon off for all visiting a local vineyard seemed mundane and ordinary compared with what we had seen and learned. While at Bordeaux, having considered for a while starting keyhole repair of groin hernias, I discussed the proposal with David Dunn and we both agreed to commence forthwith and speak regularly to discuss any difficulties and hopefully advance our techniques. We had been able to watch videos of the operation while at Bordeaux.

The meeting not only fully confirmed what many of us had begun to envisage, which is that most abdominal operations performed open would soon be done by keyhole procedures. The meeting also showed in the exhibition marquee a rapid advance in the development of instrument technology. The makers came, demonstrated and listened to what we required and went away to implement our suggestions. As an after thought the wine tasted good!

It was in 1992, having done one hundred and four keyhole removals of the gall bladder that I had my first major problem. I had to attend a committee meeting at lunch time while in the middle of an operating list and asked my assistant who had attended over fifty cholecystectomies to begin the next operation on a lady in her early fifties. I arrived and found he had displayed the anatomy and was ready to divide the blood supply and duct to the gall bladder. I made the error of taking his word for the status of the operation and, with my mind still to some extent on the content of the meeting, proceeded to divide the duct without further checking. To my horror, I found I had divided the main duct draining bile from the liver to the upper part of the small intestine, a situation, unless corrected, life threatening in the not so distant future. The operation required converting to an open procedure.

On defining the anatomy the bile duct was only of small calibre not dissimilar to that of the duct I had been attempting to divide, hence the confusion. To be able to join the two ends of the duct was a surgical challenge, as the diameter of the duct was no more than three to four mms. I was working at the bottom of a hole some ten centimetres deep and had to achieve the union without narrowing the duct or else bile flow would be obstructed causing jaundice and again long term problems. I identified the anatomy and found, as very little of the duct had been inadvertently removed, that I could approximate the ends without undue tension. I proceeded to put in a continuous posterior stitch leaving the anterior part of the tube to be closed with interrupted stitches. Before doing this, I inserted a T-tube into the bile duct on the liver side of the join and threaded one arm up towards the liver and the other through the join into the distal limb of the duct to act as an internal splint and hopefully prevent any significant narrowing. The join was then completed, whereupon the wound was closed with drainage and the operation finished some two hours plus after identifying the problem.

Post-operatively, I told the patient exactly what had happened and that I was hopeful all would be well but that she would need to have the T-tube in place for six weeks. She went on to make a good recovery and after six weeks had dye inserted into the tube and X-rays taken. These showed all to be well and I was able to remove the T-tube. I followed her up for five years when all remaining well I was able to discharge.

Once more I was grateful for the teaching and skills I had learned over the years which enabled me to overcome this life threatening situation.

After the meeting in Bordeaux, upon my return, I operated the following week on a man of sixty-eight with a left groin hernia. Being the first patient, he was kept in overnight when all being well the following morning he was discharged home. His was the first that I did and throughout the following years I kept records of what procedure I had done in each case including the size of the mesh inserted and how it was fixed. The common groin hernia results because of weakness in the musculature of the abdominal wall when a bulge, often occurring at a time of straining or strenuous exercise, can develop. This can balloon out and be filled with gut prolapsing through the neck of the opening into the sac. On occasions this can get impacted cutting off the blood supply to the contents and causing strangulation and gangrene of the contents, hence the need for repair. At operation, the contents of the sac are reduced back into the abdomen. The sac is then either itself reduced or transected before a polypropylene mesh is inserted between the abdominal cavity lining, the peritoneum, and the overlying muscles to cover the defect and act as a barrier to further herniation. To complete the procedure the mesh is stapled in place and the abdominal cavity lining repaired. Repairing a hernia by keyhole surgery ensures, unlike with open operation, that no muscle in the abdominal wall is cut and recovery therefore from operation is much quicker. I spoke regularly with David Dunn in Cambridge and we soon appreciated the shortcomings of the earlier operations. The mesh inserted we more than doubled in size and this lowered the recurrence rate from that seen in most open operations to less than one percent. By transecting large sacs at their neck rather than trying to reduce them, lowered post-operative groin bruising. Advances in instrumentation meant that smaller holes to gain access to the abdominal cavity resulted in trouble free healing, rather than leaving a potential weakness in the abdominal wall with the occasional development of a hernia at one of these sites.

In 1993, while at a SMIGS meeting I was asked by a company in the exhibition arena to come and look through special glasses which would enable me to see in 3D rather than 2D as one supposedly had with the flat screen. I tried the glasses on and said after completing a few manoeuvres that I thought I could manage just as well without wearing them. I did even better without the glasses whereupon the company representative turned to me and said I had a somewhat unusual eyesight similar to that of airline pilots who were able to put

3D on to 2D. I learned that approximately twelve percent of the population had this ability. In retrospect I realized that this may have been the reason for being told at the meeting at St. Mary's that I seemed to have a natural ability, or perhaps gift, for these procedures and should take them up.

By October 1993, word had got around Shrewsbury that I was repairing hernias using a keyhole approach and that recovery to full activities was far quicker than with open operations. Sportsmen's hernias were at that time going to a surgeon in London and the eponymous name for the condition was Gilmore's groin. He was seeing, in particular, footballers who were complaining of severe discomfort in the groin coming on after about seventy minutes playing leading to them being substituted. Within a couple of days the discomfort would have settled and they would be back in training and ready for the next game, whereupon the whole scenario would repeat itself and again they would need substitution. If they took a longer time, off once more the symptoms would return when back to full activity. Gilmore identified that the deep muscles in the abdominal wall were being stretched and tearing although there were no obvious physical signs on examination. He did an open operation and buttressed the layers as for an open hernia procedure. Following his operation, sportsmen would take around twelve weeks or more to get back to full activity when they would be found to be cured of their symptoms.

In the middle of October, I was phoned by the medical officer of Shrewsbury Town FC and asked if I would see one of his footballers who he thought had this condition. I saw the thirty-two year old player who gave a history much as above. On examination all appeared normal. The history was very convincing and somewhat sceptically, as he had tried all non-operative treatments, I agreed to have a look inside. At operation and on looking inside, I found a bulge in the abdominal wall on the side of the patient's symptoms, at the site where a direct hernia would appear. Seeing this abnormality which was not present on the other side I interpreted it as the beginning of a hernia and proceeded to repair placing a mesh in situ over the defect. The sportsman was discharged the following day and had begun to get back to light training when I reviewed two weeks later. I allowed him back to full training and twenty-three days after operation he played ninety minutes in a League match with no ill effects. The Chairman of the FC was delighted with the result having a player back in action after less than four weeks as opposed to twelve with the open operation. He rang around fellow Chairmen at other clubs stating what had been achieved. This led me to have a rapidly expanding practice with sportsmen, a number of who were household names.

As time went by, I devised a physical test whereby I was able to stress the groin muscles and could reproduce, with pressure applied in the area of concern, the symptoms of which they complained. This enabled me to differentiate this condition from other pathologies in the groin region. I soon found on looking in the groin region that there could be a bulge on the contra-lateral side which was symptomless. At first I ignored this as there were no problems or symptoms. However, it was not long before sportsmen returned with symptoms on the other side. Two of the first eight that I operated on, who had a symptomless bulge in the other groin, returned within months to have the other side done. I kept a video recording of operations on all sportsmen and when the patient returned with symptoms on the other side I was able to check back and see the original state of affairs. I rapidly learned that if I found a defect on the normal side, this should be repaired at the first sitting. In sportsmen with a "sports hernia", when examined with the laparoscope, more than fifty percent had a bulge on the other side. In most this was symptomless. By repairing both sides at the first sitting, the sportsman returned to full activities in the same time that he would with a unilateral repair. This meant that treatment could be over in less than four weeks with a keyhole operation as opposed to the open operation where, as there was no way to check the opposite side, there was the potential for more than twenty four weeks out of sport: costly for the sportsman and costly for his club. I had a number of patients, who had had an open operation earlier in London, who came to me to have the second side done. Getting back to playing inside one month was welcomed by both the sportsman and his club. Not all was plain sailing and one day I was brought back down to earth with a bump. I was operating on a thin lightly built teenager with strong abdominal muscles. The abdominal cavity was filled with gas in the normal way and then I went to insert the trocar and cannula for the camera. I met a lot of resistance which, with extra pressure being applied, suddenly gave way allowing the sharp pointed instrument before I regained control, rapid access to the abdominal cavity. On passing the camera I saw a red-out indicating rapid blood loss. I sounded the alarm bells as the patient's life was very much at risk. The team, the anaesthetist and the theatre sister, responded magnificently while my adrenaline levels upped to maximum. In the blink of an eye additional instruments appeared, while my anaesthetist applied his skills to maintain the patient's blood pressure pending the arrival of blood for an emergency transfusion. I rapidly opened the abdomen and found a rip in the main pelvic vein returning blood from the leg to the heart. With considerable difficulty I managed to stop the bleeding but not before a blood loss of three to four pints.

With control of the situation and feeling somewhat chastened, I halted to enable all to recover and to fully assess the damage. The rip was going to be difficult to repair because of access to the laceration in its entirety. I struggled, for a while without getting anywhere, before I realized that I needed another pair of skilled hands. Luckily Robert Hurlow, a colleague and also trained as a vascular surgeon, was available and with his arrival and skilled assistance I was able to complete the repair. Before closing, and after all the drama of the previous two to three hours I remembered to repair the defect in the groin on this occasion by open surgery. The patient was fit for discharge on the fourth day and went on to make a good recovery. Sadly he did not continue with football. This salutary experience brought home once more to me, however good one was technically there was always the potential for disaster round the corner even in the apparently most straightforward case. As a surgeon one always had this threat at the back of one's mind and reinforced the importance that when advising a patient to have an operation you were setting out to improve their well-being and quality of life.

By chance this was the last occasion that I needed to enter the abdomen blind as that very week I was shown a new instrument which enabled me to enter the abdominal cavity using direct vision.

The sharp pointed trocar was replaced by a cutter blade advancing 1.5 mm each time when activated and whose progress through the abdominal wall one could follow on the screen until a soft entry to the abdominal cavity was achieved. This Visiport overnight dramatically lowered the risk to the patient when entering the abdomen.

The benefits of keyhole surgery continued to be demonstrated: not only was the procedure virtually pain free requiring minimal analgesia but one had the ability to diagnose a weakness on the asymptomatic side and repair it at the first operation rather than necessitating a second operation with the inherent risks involved. Advances in surgical technique and in instrument technology improved the results to such an extent that the risk of complications became minimal and certainly less than for open operation.

There was always going to be one. A thirty-eight year old footballer playing in the lowest football league and nearing the end of his career admitted sheepishly, when seen at the follow up clinic some seventeen days post-operatively, that he had played a full game in a league match fourteen days after operation. Happily there were no ill effects and all was sound.

Chapter 14

Spreading the Keyhole Gospel

A meeting of SMIGS was held in the Spring of 1994 at Harrogate, to which the BSSG members were invited. At this meeting, it was agreed that we should unite under one roof and thus the Association of Laparoscopic Surgeons of Great Britain and Ireland (ALSGBI) was born. Lord McColl, Professor of Surgery at Guys Hospital, was elected as first President. It was agreed that the format of the meetings should remain as it had previously done for SMIGS; that is two meetings annually. The first meeting would be held under the umbrella of the Association of Surgeons meeting in the Spring, while the second in the Autumn would be in the vicinity of the surgeon chosen and with the facilities available to host the meeting. At this meeting live operations would be performed and relayed to the main auditorium. At the first autumn meeting we were then shown a video of an Italian surgeon removing a section of the large bowel for cancer and reconstituting the integrity of the bowel by doing a stapled anastomosis. It was then back to Shrewsbury to continue with the evolution of keyhole surgery.

I was now aware that virtually every operation that I had previously done open would be possible using a laparoscopic technique.

During the summer of that year, I was asked to see a female patient in her mid thirties who worked overseas for the Foreign Office and who was home on leave. She was complaining of losing fresh blood on her stool at defecation. On examination she was attractive and appeared fit. I could find no abnormality on general examination or in her abdomen. When I examined her rectally I could just tip what appeared to be a tumour. I next passed a sigmoidoscope and confirmed this finding. A biopsy was taken and the specimen sent to the laboratory for analysis. This confirmed her to have a cancer which appeared well differentiated and, therefore, probably slow growing and not too aggressive. After other tests to ensure that as far as possible the tumour was confined to the bowel and had not spread elsewhere, I recommended that I should operate and remove the cancer.

I next proceeded to discuss the operation with her which I had done open many times. The tumour was high enough up the rectum to be able to resect and then join the proximal bowel to the distal to re-establish continuity. I then told her about keyhole surgery and that I thought her an ideal candidate for this procedure. Although I had not done this operation laparoscopically before, I felt confident that it would be possible and that it would, if successful, very much shorten her recovery time. Should any problems arise, I could convert to an open operation, in which case she would expect to have a longer recovery period. After some moments thought and one or two more minor questions, she agreed to undergo the keyhole operation.

On the day of operation and in the presence of an instrument maker rep, I checked and confirmed that I had all the tools I would need. Air was put into the abdominal cavity and five port holes created. On inspection of the abdominal cavity I confirmed the presence of the tumour in the lower bowel and found no evidence of distal spread of her cancer. I proceeded to mobilize the distal colon and upper rectum, first securing the blood supply to prevent any possible tumour cells being displaced and passing to the liver where they could give rise to secondary deposits. The bowel was then freed and stapled some three inches above the tumour and a similar distance below having first washed out the distal bowel to destroy any possibly detached cancerous cells. All this took a considerable time especially the stapling while I applied the instrument at the correct angle. Once the stapled bowel was divided, the free specimen was bagged to prevent contamination while being delivered to the exterior. This was done through a hole no larger that that needed to deliver an appendix. Having closed the small hole in the abdominal wall and re-inflated the abdominal cavity, a stapling device was passed through the anus into the rectum and returning to the camera the two ends of the bowel were approximated and the stapling device engaged. This was then applied and continuity of the bowel reestablished. The whole operation from start to finish had taken just under five hours.

Post-operatively the patient made an excellent recovery and was fit for discharge after four days, the normal time to discharge for an open operation being ten to twelve days. When the specimen was examined in the laboratory, the cancer was confirmed to be confined to the bowel and to have a favourable appearance. She returned to her job within two months and came to see me again four years later when back in this country. She was very well with minimal scarring on the abdomen and no evidence of any further problems. With regard to this cancer, she could be regarded as cured but was warned that she should have regular checks throughout life as there was the possibility of her developing another cancer in a different part of the large bowel.

To train other surgeons in laparoscopic techniques, the ALSGBI organized teaching sessions, with the help of instrument makers, to run courses. On these some ten surgeons would attend together with two trainers when we would utilize training facilities in Paris and another just outside Hamburg. These had to be run outside the country because of the laws in the UK preventing vivisection on anaesthetised animals. In Paris, the establishment was near Versailles. There we would work most of the day first practicing with instruments and simulators before going on to live vivisection. These days were pleasant but demanding. Usually in the evening, the group was entertained with a trip on the river Seine in a Bateau Mouche enjoying a good meal and the sights of Paris from the river. In Hamburg, the facility was in the country near a golf course which offered the opportunity for a walk and some exercise before once more getting down to the serious business of the trip. These occasions were also a good opportunity for the trainers to discuss between themselves advances in keyhole surgery and what we might attempt next. All the time we were seeking to undertake procedures where the greatest benefit was for the patient. Recovery times from operations were rapid and almost pain free because no muscle was cut gaining access.

It was while on a Hamburg trip that we saw for the first time a built in Sat Nav in action in a car, which reminded us that while surgery was moving forward so was technology in other fields. After two nights away, we would return to the UK to resume our normal activities.

As time went by, we started training the next generation of surgeons with these techniques. First, we would introduce them to simulators when they would practice, with direct vision, using instruments to move objects from one point to another. They then moved on to using a telescope and camera when they would appreciate the difficulty of performing a Three-Dimensional procedure looking at a two-D screen. The operation that trainees were most acquainted with was appendicectomy. They would assist one at first and then, when appearing to have a grasp of the operation would be phased in, while being mentored until they could undertake the full task. Once I was confident of their abilities they would be permitted to carry out the operation but to call me if they encountered any problem.

One day during the summer of 1997, I was referred a boy of seventeen from the local school who complained of abdominal pain for twelve hours plus. This had initially been central around the umbilicus and gripey in nature. After a few hours the pain had changed in nature and moved to the right lower abdomen where it had become constant and sore. He was otherwise well. On examination, he had a mild pyrexia and a somewhat furred tongue but generally appeared fit. In the abdomen there was marked soreness and guarding against pressure in the right iliac fossa and when the hand was removed, a surge in the pain. The history and findings were classical for acute appendicitis. I took him to theatre and performed a keyhole operation when I removed an inflamed appendix.

He was discharged the next day. He was a sportsman and he told me that the school were in the middle of cricket house matches and that he was an integral part of his team. Four days after operation he played in a match without any ill effects, thus bringing home to me once more the benefits of keyhole surgery. Should he have had an open operation it would have been ten to fourteen days before he could have returned.

The West Midlands Surgical Society met twice a year and was run by a President elected annually, a secretary and a treasurer. Overseas meetings were held from time to time instead of one of the home meetings. In 1997, I was President and this happened to be a year for an away meeting. Arrangements were made to go to Barcelona. These away meetings gave an opportunity to get to know other surgeons in the region, many of whom like me had trained elsewhere. Four days before we were due to leave, Mary slipped on a tennis court tearing ligaments in her knee. She was obviously not fit to travel and but for being President, I would have cancelled. With considerable reluctance I proceeded on the trip, earning myself a number of black marks. At the meeting some papers were read but we also had time to explore the city, the most notable sight being the Sagrada Familia, a minor basilica designed by Antonio Gaudi and still being constructed. The party gelled well and certainly one got to know many colleagues in the Region who had previously only been names.

In the autumn, I hosted the meeting at the Albrighton Hotel near Shrewsbury which had the facilities for a conference numbering around eighty. Papers were read and then I spoke with great enthusiasm about laparoscopic surgery, not too many other surgeons in the region having as yet taken it up. I got totally carried away and overran my allotted time giving what I judged to be one of the worst papers of my career. I tried to pack in far too much detail rather than whetting the appetite with a view to coming back another day.

Around this time, I was asked, in my capacity as surgical tutor, if I could go to Pakistan to select two trainees to come to Shrewsbury for further training. Professor Zafa, who was President of the College of Physicians and Surgeons Pakistan (CPSP), was to be our host and asked that my wife accompany me. The day of departure arrived and we checked in to Manchester Airport to travel club class overnight to Islamabad before connecting with a flight to Lahore. About twenty minutes before we were due to board we were approached by a staff member of British Airways to say that she regretted that there were no club class seats remaining on the plane which was coming from London. Mary said she would not travel overnight economy class when we had a busy schedule at the other end, and was threatening to return home. A short time later the BA rep returned and asked us if we would mind being upgraded to first class. Reluctantly we agreed!

Having transferred to the connecting flight, we arrived in Lahore to be met by Professor Zafa and his senior trainee. We were taken to the hotel and after settling went for a walk through Lahore with our hosts. During this, the professor explained to us that at the time of partition there were only four doctors in Pakistan. The CPSP was founded only in 1962. We were informed that we would visit the main Hospital and Medical School where I would go on a ward round before giving a lecture. We were told that in addition to visiting Lahore we would also travel to Multan to visit another institute. At the hospital, I interviewed two trainees who had been put forward to come to England for further training. Both had good academic records and spoke good English. I was happy to recommend that they could come to Shrewsbury to join our training program. Both arrived in due course and were successful in their appointments. While being taken around, we saw students having lectures in the open air. In the wards, much of the basic care was undertaken by relatives who would also bring food in for the patient. After the first lecture, having already agreed to give four lectures, I was asked if I could give an additional three. Back in the hotel, I rang room service to get a beer. I was asked my nationality by the receptionist before she said it would arrive shortly. After a short while there was a knock on the door. On opening it I was handed a plastic bag, the waiter passing it over as if it had toxic contents. When I went down to reception to check on details, I was surprised when the lift door opened to see the Duke of Edinburgh. Our hosts had obviously placed us in the right hotel!

After two days we left the hotel to go to Multan. At the airport we were introduced to the class system. If you had twenty or more points you were given privileges such as using the airport lounge. Consultants had this status while trainees did not. Being a guest of the Professor, we went to the lounge while his Senior Trainee was not yet elevated to the necessary position. Boarding the plane was an experience. The plane had obviously had many hours of flying time behind it and was almost certainly second-hand. Passengers boarded with unusual articles of luggage the most notable being a cage containing hens. On leaving the terminal the plane, confirming its age, rattled towards the runway for take off. Prior to take off, the pilot over the intercom prayed to Allah requesting a safe journey. After a short flight we landed at our destination.

Multan was an old town with narrow streets, which were unpaved, and had little evidence of utilities. We were put up in the Hospital grounds and once more I was conducted on a ward round before giving lectures. At the time it was interesting to contrast the patient population compared with what we had in the UK. After the lectures, a reception was held which proved somewhat of an embarrassment to the Professor. We formed a reception line and the hospital staffs were introduced to us shaking our hands. When they got to Mary a number refused to shake her hand. The Professor apologized to us saying that these were Islamic Fundamentalists. Luckily, there were no hard feelings in the reception which followed.

After two days in Multan we returned to Lahore in another aged plane, the prayer to Allah again being extolled. A safe journey behind us, I gave two more lectures before we said goodbye to our hosts and set out on a short holiday in India.

On arrival in Delhi, we were greeted by a tour representative who welcomed us, realizing that we had just left a dry country, with a bottle of sparkling wine. We were conducted to our hotel and soon became aware of the extremes of life in the city. There were the very rich, living in splendour, and outside the walls of our hotel, among many other sites, the very poor living in shanty conditions and in squalor. In between was a middle class that was rapidly expanding. During the first days, we were shown some of the sights of Delhi. There is a major British influence with magnificent streets and parliament buildings designed by Lutyens. We discussed British rule and its achievements on a number of occasions. The consensus was that the British had left behind much that was good including the railway system. While wandering around we were warned not to give to beggars of all ages or else we would be surrounded by a whole hoard, putting ourselves at risk.

We next moved on to explore the Golden Triangle and arrived at Agra where we stayed for two nights to appreciate the beauty and splendour of the Taj Mahal. Our guide took us at different times of the day so that we could appreciate the change in colour, as the sun journeyed from dawn to dusk. We could both understand why it is numbered amongst the Seven Wonders of the World.

Continuing our journey we next moved on to Jaipur or the Pink City, the capital of Rajasthan State, where we visited the Maharajah's palace. This, now part hotel, was truly opulent and contained many exotic treasures. From there, we went on to visit the Amber Fort, a World heritage site, arriving at the entrance by way of an elephant ride. In addition to a tour of the fort we were able to appreciate the wonderful view over the Pink City.

Our next stop was at Udaipur where we stayed in the famous Lake Palace hotel, situated in Lake Pichola, and reached by boat. Until the 1970s, this lake had contained crocodiles but these died during the drought of that time. These were not replaced following the return of the rains when the lake refilled. There was much to enjoy in Udaipur, including the City Palace owned by the ruling Maharaja, the buildings and the shops containing Indian crafts.

After Udaipur, we returned to Delhi surviving the whole trip in our chauffeured Morris Oxford, the standard of driving during our journey giving rise to one or two hair-raising incidents none of which were due to our driver.

After returning to Delhi we connected with our BA flight home, looking forward to some western food, having survived and enjoyed curries throughout our trip, only to find that the plane had re-victualled in Calcutta so once again it was curry.

Shortly after returning to work I was asked by the Physicians to look at a lady in her mid-seventies. Her stomach had migrated through a hole in her diaphragm into her chest, a para-oesophageal hernia, and was causing her respiratory distress and difficulty in swallowing her food. She was considered unfit for an open operation and I was consulted to see if I could remedy the situation by a laparoscopic procedure. The patient was anxious for something to be done knowing that, without treatment, her days were numbered. After due discussion, I undertook to go ahead. For some time I had been operating, using a keyhole approach, on patients where the oesophageal sphincter and proximal stomach had herniated into the chest through a stretched hiatus giving severe heartburn. These had been cured by drawing the lower gullet back below the diaphragm and fixing the stomach with stitches to the periphery of the diaphragm and so preventing further migration.

At operation, I used the same approach as for the heartburn patients. I found that more than half the stomach had passed through the hole in the diaphragm and was now in the chest. The lower end of the gullet was situated as normal below the hiatus. I was able to reduce the stomach back into abdominal cavity, a few snips being required here and there to divide adhesions. Once the stomach was fully reduced, I found the hole in the diaphragm to be about the size of a billiard ball with firm margins which prevented closure, without tension, by direct suture. There was no tissue within the abdomen that I could use to plug the hole and I therefore decided to use the polypropylene mesh that I used for groin hernias. I fashioned this externally and then passed it into the abdomen. Once I had achieved the correct size I stapled it to the margins of the hole fixing the area adjacent to the gullet with sutures. I then tacked the anterior surface of the stomach to the periphery of the diaphragm as for a heartburn operation, so as to prevent other abdominal contents riding up and becoming adherent to the mesh. The operation was then completed. Post-operatively the patient did well and was cured of her symptoms. Not having a large abdominal incision enabled her to breathe without the discomfort and distress that she would have had following an open operation, and enabled her survival which would have been threatened with an open procedure.

I was asked subsequently to operate on a further five patients with this condition, two from my own hospital and three as an outside referral. All operations went well and the patients were cured of their symptoms. Three years later, I reviewed the first patient operated on while she was in hospital for another condition. She had remained well and symptom-free in the interim.

Not only was I helping to train English surgeons on overseas courses but I visited some other hospitals in the region and demonstrated keyhole procedures for gall bladder removal and hernia repair. There was also a responsibility towards the trainee surgeons on one's firm and helping to advance their careers.

Chapter 15

Towards Retirement

In addition to participating in patient care in outpatients, on the wards and in the operating theatre it is necessary for trainees to join in any research that may be under way and to be authors or co – authors of any publications. Publish or perish!

Between 1992 and 2000 I kept a detailed record of all patients operated on for groin hernia. This enabled the trainees to participate in or write some seven papers of varying interest. The contents of these demonstrated a method of inserting trocars through the abdominal wall to avoid blood vessels and thus preventing abdominal wall bruising, to a controlled trial as to what action should be taken when repairing a hernia if a defect or bulge, but not as yet a hernia, was seen on the contra lateral side. The latter was the subject of a randomized controlled trial in which the defect was repaired in one arm, while in the other the defect was noted and left in situ. It was not long before those which were not repaired were returning with a hernia on the side of the defect, necessitating a second operation. These findings fully justified taking the decision to repair the defect at the time of the first operation, thus saving the patient further surgery and, for the hospital, creating theatre time for another patient. Other communications outlined the improvements made in the keyhole repair of hernias. These showed, early on, a complication rate of around twelve percent, most of which were minor, and which were similar to those seen in general with open repairs. As the technique was refined, for example, by inserting a larger mesh and measures were taken to reduce, for example bruising, the complication rate diminished to be less than that for open operation and to be equal to that achieved by surgeons considered experts in the open operation. The recurrence rate of the single-sided hernias was similar to that done by experts but superior for bilateral and recurrent hernias. The latter, when repaired open by experts, had a complication rate of around twenty percent mainly groin and/or referred pain down the leg but also further recurrence. When repaired using a keyhole technique, the complication rate was the same as that for primary operations.

During my career, administrators began to take increasingly more action as to who we should treat and when. Extra administrative staff were appointed to ensure Government dictates were upheld. Usually an administrator would not take individual action but would take a problem back to a committee so that if a wrong decision was made no one individual was responsible. This was a very different situation to that of myself and my surgical colleagues where as consultants in most cases we had to make instant decisions often in life-threatening situations. The buck stopped here!

Courses would be run in the regions for administrators who after attending would, on occasions, call a meeting of consultants to try and institute some method that they had heard of with a view to trying to save on bed occupancy and improve throughput. As consultants with overlong waiting lists, over the years we had tried many manoeuvres to try and solve this problem. Invariably, after being addressed, we would either say that we were already using that procedure or else we had tried it in the past and had abandoned it as useless. Another major irritation arose when I was told by the admissions clerk that I had to take a patient off the next admission list and substitute it with one who had been on the waiting list for some six years. This patient had a cosmetic lump which required removal as much as anything for vanity. The patient that I was asked to postpone was getting severe biliary pain at least once a week, necessitating the GP being called out to administer analgesics on each occasion. This patient had a young family and life was becoming impossible, not knowing when the next attack would arrive. I kept a list of deserving patients either resulting from my seeing them in outpatients or on a domiciliary visit or because the GP had requested that the admission be expedited. This patient was on that list. I told the clerk that if the patient I had selected was substituted I would cancel the whole list. Furthermore I said that if this problem arose again I would take the same action. I also made the point that patients would be selected on the basis of clinical urgency and not on account of Government demands to implement and massage waiting times. I am sure that my views were passed up the administrative ladder but I never had a come back. By this time my retirement was not many years away but I thought actions like this would be very difficult for a newly appointed consultant who would find it almost impossible not to do what he was told.

Despite these irritations, I continued with a busy practice. I remained doing my on call rota being liable to be called out at any hour of the day or night when on duty.

On one occasion I was called in to see a farmer who had been gored by a bull in the lower abdomen. He was in a bad way and following resuscitation needed to go urgently to theatre. On opening his abdomen through a low incision, I found damage to the large bowel, a hole in the bladder and a rip in an iliac vein which conducts blood from the lower limb back towards the heart. It was then that I was grateful for my very general surgical training, consultants by then, on appointment, not having the breadth of training that I had, but tending to specialize in a particular field such as urology, colo-proctology or vascular surgery. I began by controlling the bleeding and suturing and repairing the hole in the vein. Once achieved, the abdominal cavity was washed and cleaned out. I next repaired the bladder and then the small hole in the large bowel. At the end of the operation a temporary catheter was placed in the bladder to allow healing. I judged that a colostomy to allow healing of the large bowel was not necessary as the hole had been small. The abdomen's apron, the omentum, was placed adjacent and fixed to the region of the hole to give added protection by walling off the area from the rest of the abdominal cavity. The patient was placed on antibiotics and made a good recovery, being fit for discharge two weeks down the line. He was told that he had been lucky to get away with this accident and in future to beware of the bull. Incidentally, without the general training that I had received, this operation would have required the presence of three surgeons in the disciplines outlined above.

During my last four years working for the NHS, I was twice invited to attend the SAGES meetings in the US, one meeting being in Chicago and the other in San Francisco, a favourite venue for me.

These meetings were vast, the number of delegates being measured in thousands rather than hundreds. Within the meeting there was a large trade exhibition where I was able to meet and discuss instrument advances with reps that I often knew. One morning, some months before attending one of these meetings, I had reps from two different companies asking me to trial a new trocar and cannula that each was beginning to market. After trialling both I said each in its own way was no advance but added if one feature from both were combined in a new instrument then that would be progress. At this meeting again sitting in a sound proof box I was told that because of patents the big company had had to take over the smaller one and that the instrument would appear shortly. A major feature of this was a small knife on the tip of the trocar with a dilator behind which stretched the tissues as the abdominal cavity was entered. When the trocar was removed the cannula was of the same size as previously used. On removal at the end of the operation the stretched tissues would resume their normal position. There was no longer a hole through which small bowel had occasionally herniated requiring a further operation. This advance helped once more to remove a possible complication. The SAGES meetings were similar to the UK Association of Surgeons meetings in that they were multi-disciplinary but much larger and once more provided an opportunity to meet again surgeons I had met previously while doing vascular surgery. These meetings were of value not only for observing the most recent technical advances and equipment, but also for seeing how other surgeons carried out differing procedures. By viewing videos, shown during their presentation, one could compare one's own manoeuvres with theirs and, on occasions, as a result refine one's own technique. There was also the opportunity to discuss afterwards, with the lecturers, why they might choose to operate in a particular way and determine if it had any merit over the way in which *I* might do the same operation. Information gained in this manner would not be available if reading a paper written by them describing the same operation. There was always much to gain from these meetings and I returned to my own work place enriched by what I had learned and stimulated to progress and advance further. As keyhole operations became more refined so complications diminished and recovery times lessened, again of benefit to the patient but also to the administration in that with lengths of hospitalization reducing there was less pressure on beds.

With these major improvements in surgical abilities I was being asked to see more elderly patients often carrying other pathology in addition to that for which they were referred to me. In these patients their quality of life was being impaired, hence the reason for referral. I worked closely with my anaesthetist who would assess the patient's general condition. Not infrequently he would say I can manage the patient under general anaesthetic for a certain time. Working with me regularly, he knew my capabilities and would just say you need to have the patient off the operating table within a set time. On a number of occasions elderly men with painful groin hernias came into this category. Only once did he turn down a patient saying the risks were too high. This patient had to be operated on under local anaesthetic. The repair was achieved but the patient in addition to his other problems then had to tolerate considerable groin discomfort during his recovery. With all the other patients that were considered, it was agreed that I could proceed. I managed to complete the operations within the set time. It was a matter of operating under a degree of pressure and on some occasions an adrenaline surge.

Towards my retirement, the European Working Time Directive came into being. Because of the limitation put on trainees' working time each week, this meant that those doing their house jobs were not always available to attend consultant and registrar ward rounds, thus missing out on teaching opportunities but worse still interrupting continuity of patient care. This tended to be worst at weekends when three or four different house officers might be responsible for the front line care of a patient during this time. With their inexperience and only caring for a patient for a maximum of twelve hours at a time, they sometimes failed to appreciate a slow deterioration in a patient's condition. By the end of a weekend a patient could have changed from being satisfactory and progressing well to a situation where they had developed a major complication and their recovery was seriously threatened. Although a consultant would have been on call and available in the hospital, his attention, which would have allowed the necessary action to be taken, might not have been drawn to this patient. Not only was house officers training and experience limited by this regulation, but also that of those nearing consultant appointment. In some hospitals, by signing a declaration, trainees could opt out of the directive, thus enhancing their training opportunities. In others opt out was not permitted, and this was undoubtedly to the detriment of patient care not only there and then, but in the long-term by limiting the operating experience of Registrars. The knock on effect of this was that when appointed as Consultants they could still be very much on the learning curve, not having the numbers of patients and experience under their belt to counter problems which would inevitably occur.

During my last year of working before enforced retirement, when on call one night, I was summoned at around one a.m. to attend the hospital where a man in his mid-seventies was being admitted with a leaking aneurysm. I gave the Registrar the necessary instructions regarding getting an abdominal X-ray, ordering blood, alerting theatre and summoning an anaesthetist. A short time later I met up with the team and examined the patient in the anaesthetic room. I found him to have a reduced blood pressure and quickly confirmed the diagnosis. The anaesthetist had gassed patients previously for me with a similar condition so knew when to transfuse the patient rapidly and when to relax the abdominal wall, a potentially dangerous moment in the operation. With the usual tension in the theatre atmosphere and an adrenaline surge in me we got under way. I was rapidly into the abdomen and confirmed the presence of the leaking aneurysm, the haemorrhage being confined and controlled by the posterior peritoneal wall of the abdominal cavity. Having delivered and bagged the gut outside the abdomen, I placed retractors to give access and then opened the posterior peritoneum. With a mixture of blunt and sharp dissection I got round the aorta above the pathological dilated part and achieved control. In a similar manner I gained distal control of the aneurysm whereupon we could all relax, the situation now being safe. I next opened the aneurysm before choosing and inserting an appropriately sized graft, a gortex tube which was then sutured in place proximally and distally to normal aorta to achieve blood tight joins. The posterior abdominal wall was then closed and the displaced intestine put back in the abdominal cavity and the wound closed. Post-operatively the patient went to ITU to be monitored and recovered. All was well for the first few hours but then the patient's condition deteriorated. We were unable to discover the reason for this before he died three days after what appeared to have been a successful operation.

At post mortem he was found to have a hole in the small bowel which leaked into the abdominal cavity giving peritonitis from which he died. In every other respect the operation had been accomplished successfully. Not surprisingly, I was very upset by this finding. The hole in the intestine almost certainly was due to a retractor injury, which I might have checked for prior to closing. This to my mind raised the question as to whether, approaching retirement, I should have been on call following a normal day's work beforehand. Since my retirement this situation has been remedied, as surgeons over the age of sixty in this hospital are no longer on rotas covering emergency night work. I remember at the time not sleeping after returning home, the body still being supercharged, and then having to cope with the next day's work. Prior to becoming around sixty years old I had found no difficulty in coping with and managing this kind of situation.

During my last few months of NHS work, knowing that the skills of the other members of the surgical staff and my replacement were not up to doing the range of laparoscopic procedures that I had undertaken, I operated to clear the waiting list of those patients. Most of my colleagues could do a keyhole removal of the gall bladder and one, in addition, was proficient at hernia repairs. Other than a keyhole appendicectomy, none of the other operations that I did was within their compass. Once a week, I did day-case keyhole hernia repairs. I was able to reduce the waiting list a little but almost as fast as the list was cleared, I was adding to it, there being constant referral of patients with this condition.

In May 2000, the time came for my retirement from the NHS. This took place over two weeks with a number of functions and the need on occasions to say a few words. I had spent twenty-six years working in Shrewsbury and had performed more than thirty thousand operations. During this time, I had seen structural changes, the largest being the completion of the Royal Shrewsbury Hospital and the building and opening of Telford Hospital. It was probably an error to have the medical services on two sites in the County as, with changes in the training of surgeons and the increased specialization, new appointees did not have the background to cover emergencies in the way that surgeons of my generation with a very general training had. It was changing from surgeons being Master and Jack of all Trades, to Master of one or at the most two. This meant that back up was required in each discipline. In the course of time the services were rationalized out of necessity and ended up with specialities being concentrated under one roof so that no longer was each discipline under both roofs. It would have been so much better if this possibility had been visualized in the seventies and one large hospital, as had been mooted, containing all disciplines had been built midway between Shrewsbury and Telford.

In addition to the structural changes there had been major advances in the equipment and instruments that were now available to us. In Radiology, in addition to X-rays, we now had CT and MRI scanners which gave far more information regarding any pathology a patient might have and reduced the incidence of having to open a patient's abdomen to assess the extent of a cancer, and whether or not it might be curable by operation. The coming of laparoscopic surgery, as has been seen, was also another major advance.

During my time the face of surgery had very much changed. When I arrived in Shrewsbury, there was a lot of surgery for peptic ulcers and the complications thereof. With the evolution of medical treatment, the need for this had largely disappeared. For the first ten years in post, I undertook basic urological procedures such as prostatectomies, trans-urethral for the smaller prostates and open for the larger ones. Kidneys were removed for cancer and stones in the renal tract extracted. After ten years, the management of these conditions was undertaken by the then appointed urologist. Operations on women for breast disease similarly came within the domain of a single surgeon. The treatment of patients with vascular problems had also moved on, much of what I did open early in my appointment now being done by interventional radiologists. Their more modern equipment enabled them to gain access to arteries, dilate narrowed areas and place stents to hold a vessel patent. With the need for surgery in certain conditions having diminished and patients living to an older age there was an increase in the numbers of patients with cancer requiring treatment. Anaesthesia had also progressed so much, so that patients early in my appointment who would have been turned down for an operation would now be considered, particularly where the quality of life was impaired. At the start of my surgical training, patients in their seventies requiring a major procedure would often have been turned down as too risky, whereas at the time of my retirement age was no barrier, rather their general condition and the quality of life after treatment. The arrival of keyhole surgery had further increased the field of those suitable for operation, abdominal pain following a lengthy incision being eradicated.

On my arrival in Shrewsbury in 1974, my predecessor had passed over the patients under his care to me by doing a ward round. In addition he introduced me to his medical staff and the nursing staff running the wards. I performed a similar function for my successor. With patients remaining in hospital after operation for a considerably shorter time due to quicker recovery from anaesthesia and early ambulation together with the arrival of keyhole surgery, the number passed over was considerably less. During my latter years there had been a considerable change in nursing protocol in that patients tended to be called by their first names. This was a habit I disliked and would not allow on formal ward rounds as, with increased familiarity, there appeared to be a drop in standards. I felt it was difficult to communicate sometimes unpleasant news if relationships were too chummy. Nursing, by this time, seemed to be centred round the desk with its computer rather than at the bedside, in some cases to the detriment of patient care. Following my handover round, I left the ward for the last time on Friday 12th. May 2000

Throughout my years in Shrewsbury, work and life in the hospital had in the main been a happy experience. My time had been much helped by having the same secretary throughout the years. Knowing how I worked, she was a great intermediary between me and GPs. Almost invariably she would know how to contact me even if I was out with a GP on a domiciliary visit. The ability to contact, of course, became much easier with arrival of the mobile phone. She was the only person able to interpret my scrawl, this not being possible even for my wife, Mary.

As I left the hospital and subsequently, I could not help but think how profligate was the NHS in not keeping me on in a training capacity which I would willingly have undertaken. In a mentoring role I could have continued helping with the training in laparoscopic techniques of the Registrars and occasionally assisting consultants wishing to expand their repertoire. I had spent ten plus years at the fore-front of this development only for the skill to be jettisoned. At that time there were only a few units in the country with this wealth of experience and in a small way it could be likened to allowing a return towards the Middle Ages!

Chapter 16

Challenges in retirement. One of my Nine Lives.

Following my retirement from the NHS, I was fortunate still to have a private practice. Having worked at high intensity during the preceding few months, going from being flat out to nothing could well have been the death of me. I had been voted on the ALSGBI Committee in the year before I retired. My term was for three years in the first instance and while I was still in touch I would complete this spell which was acceptable. I would not seek re-election as by then, I judged, I would be remote from everyday hospital practice.

I decided to pursue two activities. During my last eighteen months of hospital practice, I had been approached to do medico-legal reports on patients suffering from complications and in occasional cases death from surgery, one of which was laparoscopic. Now that I had the time I was able to enlarge this activity. The other field I chose to explore was to collate my work on keyhole hernia operations. I was still operating on these and in eight years had carried out some two thousand six hundred repairs on one thousand nine hundred patients. I had kept accurate records of each operation and, in many cases, had filmed the procedure. All repairs on sportsmen, in particular, were videoed.

In 1999, as the family was now mainly away, we moved house finding a property in Cardington, in the country and adjacent to Caer Caradoc, south of Shrewsbury and off the main road between there and Ludlow. Of particular joy to Mary and myself was our youngest daughter Sarah's decision to enter Medical School. She would become the fourth generation to enter the Medical Profession. At that time there were two choices. In the first she could follow the same path that I had done doing the basic sciences, anatomy, physiology and bio-chemistry with no clinical input until after second MB or, a choice only recently available, she could do these combined with exposure to patients and illness from day one. After interviews and offers at colleges in both disciplines she chose the latter and went to Imperial in London.

In early retirement there was a marked change in the pace of life, giving me an opportunity to follow pursuits and pastimes that had not been possible when working. I was able to help in getting our new house and garden established so that it was as we wanted it. I was able to play golf during the week and in inter club seniors' matches, and go for walks with the dogs in the countryside surrounding our dwelling. Work now occupied about two days a week, doing an outpatients and an operating session in the private sector and reviewing patients involved in medico-legal cases. I had to go through their hospital records with a fine-tooth comb looking for inconsistencies or comments that might indicate where a problem arose. I then had to interview and examine the patient, where possible, before writing a report and venturing an opinion as to whether the patient's treatment was up to standard. In the event of an ongoing problem, it was necessary to state what the long-term outcome might be and whether further surgery might be required to rectify the situation. This work could be stimulating and to me appeared to be a form of detective work, looking to see in some cases whether blame could be apportioned in any area.

I next started to collect and analyse all the data I had on keyhole hernia repairs. The first repair dated back to 1992. Early on, I spoke regularly with David Dunn in Cambridge, and the method metamorphosed during the next five years to a standard procedure. During this time, there was also an evolution of surgical instruments from a level of being relatively primitive to becoming more developed and, with the aid of feedback from the likes of myself, fit for purpose. As techniques evolved, one was able to enter the abdominal cavity using an optical trocar (Visiport) which virtually removed the risk of damage to intra-abdominal contents. Once inside, one gained the means to trans-illuminate the abdominal wall and so avoid puncturing abdominal wall vessels when placing lateral access ports, and with the development of dilating trocars lowering the risk of a port site hernia following surgery. Access to the abdominal cavity thus became standardized and with it the removal of a number of complications, some of which had required further operation either immediately or at a later date.

During the period of instrument development, the operation in the groin also evolved. Initially a small six by five centimetres mesh was used. At the very beginning this was not fixed when introduced to the pre-peritoneal space but because of recurrences was then stapled in place. As time passed the size of the mesh inserted was further enlarged first to eleven by six centimetres and then finally to fifteen by ten centimetres. where it remained. The treatment of the hernial sac also changed. At first, these were removed in total but this provoked post-operatively marked bruising in the groin, not a pleasant sight. The sac was then transected at its neck leaving the distal portion in situ. This cured the bruising but, on occasion, gave rise to a transient collection of fluid in the remnant which nearly always absorbed spontaneously. Following a randomized controlled trial, the treatment of a defect found coincidentally in the "normal" groin at operation for an apparent single side hernia also changed, a repair being done.

All these changes enabled a new approach to hernia repair to become established, giving long-term results as good as the best achieved with open repair but considerably better than those for open repair of a recurrent hernia. From the patient's viewpoint, recovery and return to normal activities was more rapid. After undergoing a keyhole operation, even as a day patient, almost fifty percent required after leaving hospital no analgesia or pain killers. A further advantage was the ability of the surgeon to assess both groins at the time of operation and repair a defect on the "normal" side thus obviating the need for a second operation.

While collating all this information, I was able to determine the longer term results of this operation and in particular those done on sportsmen. Longer term follow up, from three months to four years, was done either by contacting the sportsman and having a telephone interview or contacting the sports physiotherapist, most of the sportsmen being professional footballers. Of 287 sportsmen who had undergone operation I was able to get details on 192. Of these I managed to get information on 176, the others having retired, were living abroad or were untraceable. 174 were symptom-free and two had minor groin discomfort not preventing them from indulging in their sport. Once the method of repair had been standardized, there had been two recurrences after 448 repairs and while I had treated a number of patients who had a single side open repair elsewhere, I had no patients returning to have a further operation where a defect had been identified on the first occasion, emphasizing the benefits of keyhole surgery both to the sportsman and his club.

When I had gathered all the information about this operation, I submitted an abstract to the Royal College of Surgeons and applied for a second Hunterian Professorship. In 2001, I heard that my application had been successful. I delivered my lecture to my peers, a challenging occasion, at the autumn meeting of the Association of Laparoscopic Surgeons held in Guildford. This appeared to be well received and I was delighted that my family was represented by my wife and Sarah our daughter who had now started as a medical student at Imperial.

Once retired, all was not work and we were able to go on extended holidays. It was also my turn to marvel and appreciate the benefits of modern surgery. Hip replacements had come a long way from those that I witnessed my father performing in the 1950s when a Judet Arthroplasty was the vogue. Not having good long-term results, this was followed by the Charnley prosthesis, developed in Wrightington Hospital, which incorporated not only new materials but also design features arrived at after considering basic physical principles such as lines of force and coefficients of friction. As time went by, the life expectancy of the Charnley replacements from normal wear and tear steadily increased from around ten years to more than twenty. No longer would a patient with incapacitating arthritis be confined to a wheelchair for mobility.

In 2002, I underwent a right hip replacement, and was told that I would need the left side doing in the not too distant future. This held out until 2009.

Later in 2002, Mary and I went with two of my brothers and their wives to India. We started our trip in Chenai. After two days there, sightseeing and visiting an open air school in session, we caught a train very early in the morning and travelled to Mysore. There we visited the magnificent palace, home to the previous rulers. We then travelled by mini bus to Ooty, previously a summer home to the British which set at 7200 ft above sea level remained cool in the hot season. There we took a trip on a local mountain steam train passing through countryside with outstanding views of valleys and vegetation and visited the botanical gardens. My brothers and I then played a round of golf on the local course where, because of the height, we were grateful to have caddies, and ball spotters. I was more than peeved on one hole to hit a good drive down the middle over a marker, the fairway being blind, while my brothers appeared to drive well into the woods on either side. After climbing the hillock we found all the balls in the centre of the fairway with the ball spotters returning from the trees on each side!

From there we journeyed into the heart of the State of Tamil Nadu to Madurai and went on a guided tour of the Meenakshi Temple where we became acquainted with the history of the Hindu religion. Tamil Nadu was one of the poorest states in India and the poverty in the population was very visible. We then journeyed on to Kerala, one of India's wealthiest states, and the difference was very visible. The people were well dressed and clean and the housing of a different standard. The streets were tidy and unsoiled and no longer did one need to look to see where you were placing your feet. Cars were aplenty. We stayed south of Kochi and visited the backwaters taking a boat trip and enjoyed the wonderful scenery, the vegetation and animal life.

The next day I was swimming in the sea with my brothers. The waves were quite powerful there being nothing between this shore and Africa. When I came to get out, I dived to ride a wave to the shore. My head banged the steeply shelved beach which I had failed to appreciate and I immediately saw stars. When I came to move my limbs nothing happened. I was paralysed and face down in the water. I tried once more to move and again nothing happened. The sea was well lit by the sun and I was looking straight at the sand beneath. My brain was functioning normally and after uttering a swear word to myself, I quickly thought I could hold my breath for a maximum of three minutes. My muscles were not moving and therefore my oxygen requirements were much reduced. Unless rescued in that time I would drown. My brothers had reached the shore and after a moment realized that I was lying facedown in the sea and floppy. They raced in to the sea and pulled me out bleeding from my forehead which had collided with the beach. One of my nine lives! John, my medical brother, quickly examined me and confirmed my paralysis from what seemed to be a whiplash type of injury. I was helped back to the hotel room and slowly began to get some feeling back in my lower limbs. It was necessary to go to hospital and an ambulance was called. While driving through Kochi the ambulance was run into by a motor bike, the bump from which I felt while lying on the stretcher. The hotel manager who was accompanying us in case of language difficulties immediately turned and said "That's good, the evil spirits which caused your accident have passed to the cyclist. You will now be fine". In the hospital, I was seen by a surgeon who had done much of his training in Cardiff. He arranged for some medication and for me to have a CT scan. It was lucky that Mary had bought all the papers with us as everything had to be paid for up front. Knowing the cost of scanners in the UK I was surprised to learn that there were four in Kochi almost certainly purchased at a price well below those in the UK. Over the next 24 hours, I began to get some movement back, particularly in the legs. It appeared that I had been lucky, just avoiding permanent paralysis. The surgeon said I was fit to travel provided I returned to the UK in business class. From Kochi we travelled south to the nearest airport from where we flew to Dubai to get a plane back to the UK. My brothers benefited from my injury also travelling home in Business class. I could not write properly for a month due to weakness in my right hand but then function steadily recovered as the bruising on the nerves receded and I was able to operate again within two months.

In 2004, we decided to move to a smaller more practically sized house. We spent some months looking at properties on the edge of villages in Shropshire but to our minds we couldn't find one suitable. The character of some villages which we might have considered ten years earlier had been changed by the addition of housing estates into which had been moved people from poorer properties in Shrewsbury. Some six months after starting our search, we went to Dorset for a weekend to attend a Nightingale nurses' reunion held by one of Mary's group of friends in Cerne Abbas. We stayed with her longest standing friend near Swanage in a property where they lived on land owned by the National Trust. She and her husband had converted a run down farm house into a beautiful dwelling, on a magnificent site on the Dorset coast. We arrived in Cerne Abbas early and had a good nosey around. It is a beautiful village with a lovely atmosphere and quaint properties, many dating back several centuries. We suddenly thought why remain in Shropshire, why not move to a new area like this. On reaching the party, we discussed with our hosts the state and supply of properties in the village. They said that houses rarely came on the open market and that if they did they were usually gobbled up rapidly. Low and behold in the Sunday paper the following morning a converted old bakery was being advertised. After a quick look and advice from a friend we bought the house and moved there some six months later. Being retired and without the benefit of school age children who help to enlarge one's social circle, we resolved to attend all clubs and societies' meetings in the village. Our new neighbours held functions to introduce us to village worthies and it was not long before we were integrated into the village life. I was soon asked to participate in the Village Fete and some six months later was made chairman for the following year. Over three years we managed to increase the takings donated to local charities by fifty percent. For fetes to be successful, fresh input is required to obtain variety and appeal so after three years I passed this over to a successor. I joined a golf club and became interested in a new village hall which was in the process of being built. This project which had only slowly progressed from conception to near completion in nine years, due to administrative problems, had been driven by a committee which was becoming tired and wanted change. Together with a retired naval officer in the village, I visited other village halls throughout West Dorset to garner information as to how they were run, what equipment was installed, what facilities they offered and what charges they made. After a few months we were asked to equip the new hall and were appointed as the nucleus of a new management committee. A successful business man was enticed to become treasurer and with the appointment of a bookings secretary we were under way. To keep hire charges low for village users, we resolved to keep the hall in good condition, and to keep facilities and equipment such as projectors fully up to date so that it would be attractive to hirers from outside the village. This policy proved correct, hire charges from users in West Dorset contributing to the well-being and maintenance of the building. Some ten years on the hall has proven to be a major village asset.

In 2007, I noticed, when playing with my old stethoscope, that I had a mild murmur over my heart which investigation showed to be due to some narrowing of the aortic valve. The pressure gradient across the valve was acceptable and I was told this might deteriorate slowly in time. I was placed on observation and told to attend for review in two years time. In 2008 prior to going to Australia to visit our youngest daughter now qualified and working doing her F2 or second year jobs, to achieve registration, I needed to get the left hip renovated. When attending for the routine pre-operative assessment, the nursing sister before giving me the green light to proceed said that I should have my heart re-assessed prior to having an anaesthetic. Although I was symptom free, the repeat echocardiogram showed that my aortic valve had further stenosed and that it was now in a critical condition and a serious risk to my life. The anaesthetist, quite rightly, turned me down for hip replacement until this was sorted. Having a friend at the Brompton hospital I was referred there and following a full range of tests was seen by the professor of surgery, a leading world expert in valve replacement, and placed on his waiting list.

I was soon admitted to hospital and prior to operation was faced with the consent form and signing this after the risks of the procedure, including death, had been pointed out to me. While I could have been frightened by this, I fully realized what the outcome without surgery would be and that with a successful operation, bearing in mind my hip was yet to be done, I would return to a near normal way of life. As I had said many times throughout my surgical career to my patients "it is the quality rather than the quantity of life that matters" and that without this operation there would be no chance of quantity. In a positive frame of mind I entered the anaesthetic room to awake some hours later in the ITU with a pig valve in situ. Soon the various drains were removed from my body and while making good progress I was once more pleased to see that measures to prevent deep vein thrombosis were in place. Once discharged, I returned to Dorchester and was referred to the cardiac rehabilitation clinic, where together with patients recovering from heart attacks and various other forms of open heart surgery, I was supervised doing graduated and increasing exercises to return my activities to near normal.

With the cardiac problems behind me I returned to Dorchester to get the hip fixed. On this occasion my pre-operative assessment was entirely satisfactory and arrangements were made for my admission to hospital. A little drama was associated with this. Late in the evening prior to my admission I was phoned and told that there were no beds available and that my admission the next day would have to be postponed. Having worked in the system myself, I decided to remain starved as for the operation. My hunch was correct because the anaesthetist, on hearing that I had been postponed, soon found a bed and rang to see if I was still starved. On hearing that I was, I was admitted and underwent replacement surgery later that day. An advantage of having previously worked in the NHS for many years and knowing the ropes and its shortcomings!!

Once more the operation proceeded without problems and following a further period of rehab I was back to full activities including golf.

When fit again, we returned to planning our trip to Australia and visiting our daughter.

We set off in December. On reaching Heathrow we joined the short queue to check in. Ahead of us, a man was having an argument with the BA rep and giving her a rough time. When it was our turn we sympathized with her for having to put up with all the agro and when my wife asked if there was any chance of an upgrade we were offered two seats in first class. My wife had resolved not to drink on the way out, but when she discovered that Bollinger was on offer she rapidly changed her mind. After a comfortable trip we arrived in Sydney and put up in a hotel for a couple of days to give our body clocks time to adjust. We then met up with our daughter who was working in the Royal North Shore hospital. She was sharing a flat overlooking Sydney harbour with a view across to the Opera House, an idyllic spot. She could walk down to ground level and swim in the harbour water and, as the building was adjacent to the Royal Sydney Yacht Squadron, she had been fortunate to meet a boat owner who had invited her to become a crew member. This she was doing in her spare time. Sydney was a sprightly city and very attractive for the young, both to settle and work in.

During my surgical career in Shrewsbury, I had worked with and helped to teach a number of aspiring Australian surgeons who were nearing the completion of their training. They were with us for a year and during this time were able to hone their surgical skills. Before leaving for Australia I had been in touch with David Pollock, my fellow consultant surgeon, and had found their contact details. One of these had organized a social evening when he had managed to arrange for six past trainees to be present. It was a wonderful evening overlooking and looking down on Sydney harbour, when we had a splendid reunion. I was delighted to learn that all but one had had successful surgical careers and we were able to reminisce on the past in Shrewsbury and the subsequent years since we had last met. The following day we had an extended trip round the upper reaches of Sydney harbour in the boat of our host of the previous evening,. Our daughter had got engaged and her fiancée joined us for Christmas. On the following day, Boxing Day, we embarked on the James Craig, a Tall Ship, and sailed out of Sydney Harbour into the sea where a pod of dolphins thirty to forty in number laid on a magnificent synchronized display, jumping out of the water and reminding us of shows put on by the Red Arrows. After this, following the start of the Sydney to Hobart sailing race, ninety-nine competing boats passed on either side of our ship, another great spectacle. The following week we visited Alice Springs before returning to Sydney to join the celebrations of New Year's Eve and see the spectacular and noisy firework display from close to. Mary and I then took in Melbourne, Adelaide and Tasmania before travelling on to South Island, New Zealand, where in addition to enjoying the wonderful scenery we met with an old colleague of mine from Shrewsbury, Graeme Kerr, a retired Gastroenterologist, who lived at Picton in the North of South Island. With him, we explored Queen Charlotte Sound in his boat and on the next day visited the adjacent Marlborough Vineyards sampling some of their produce. Before leaving New Zealand we went south and joined a boat trip for twenty-four hours on Doubtful Sound to see some of the most beautiful scenery we had ever visualized. Towards the end of our trip we took a helicopter flight up Fox Glacier where again the views were exhilarating. We then returned to Australia and visited friends before returning to the UK.

I had been troubled for a number of years with arthritis in my left knee which had resulted from a soccer injury aged eleven and which had caused my knee to be misshapen (Genu recurvatum). I had seen colleagues previously who had said that technology was not yet up to coping with a knee of this configuration and to persevere for the time being. By 2012 technology had moved forward and I saw a surgeon in Exeter who, in addition to a surgical background, also had an engineering degree. He was highly recommended to me by an orthopaedic surgeon in Shrewsbury who had been one of his mentors while training. At consultation, he said that the surgery would have to be undertaken in his main theatre where computer studies, because of the complex shape of the knee, could be undertaken. In view of the severity of the symptoms, I duly underwent a total knee replacement. Once more I had a successful operation and made an excellent recovery. Now having a normal looking and pain-free knee, no longer am I embarrassed to expose it to public view.

Yet again I returned to full activities returning to playing golf and participating in my duties as a member of the Seniors Golf Club committee, where I was responsible for team selections for matches. I continued playing club bridge and remained as chairman of the Village Hall committee.

I had been fortunate to be able, with the help of modern surgery, to continue with a normal way of life and activities, which if of my parents' generation would not have been possible. Throughout all these adventures I had been ably supported by my wife who encouraged me to undertake rehab and regain my mobility and independence, and without whose enthusiasm and drive my recoveries would have been considerably delayed.

Chapter 17

Guillain-Barré Syndrome

Christmas 2015 was spent at home when we were joined by family, our oldest daughter Katie, and youngest, Sarah, the latter with her husband and her two children. Christmas day was exciting especially for our grandson who was three and a half. Our granddaughter, one and a bit, was opening her eyes to this celebration for the first time. After a joyous day devoted to the grandchildren, my oldest brother, Peter, and two of his sons joined us on Boxing Day for lunch. It was an opportunity to catch up on the lives and careers of my nephews.

Next day I went with my daughters and family to see a travelling Pantomime, Cinderella, before Sarah and her family returned home. It had been wonderful seeing them and the progress they were making. That evening we were joined by our son who had flown in from Moscow, where the date of religious festivals follows the Julian calendar. Next day I was laid low with a chesty cough which persisted with varying degrees of intensity for three weeks. After four days I judged myself no longer infective to others so my son drove me up to one of my brother's who lived near Cheltenham, to celebrate New Year. Sadly Mary had succumbed to the bug and had to remain at home. Being a party man I overdid the celebrations and saw in more of the New Year than was prudent. The following day, my cough relapsed but I got through it, albeit with fits of coughing, before returning home to retire from the festivities. I remained low for two weeks before returning to my normal routines and playing golf once more. My first game after three weeks lay off was abysmal and my partner and I were knocked out of a winter competition. I played golf again on the Friday when my game was somewhat improved

I went to bed normally that night but on waking, in the morning, found I had some weakness in my right arm. I thought I must have slept in an odd position and that this when I mobilized would wear off later in the day. I got quite a shock when I got up and found I also had some weakness in both legs and walking required some effort and concentration. I decided to say nothing to Mary and to see how I went, hopefully improving. As the day progressed I became aware that I was getting weaker. Being an avid follower of cricket I sat most of the day watching an ODI from South Africa, hiding my progressively advancing weakness. Come late afternoon as we were hosting some friends to supper and bridge I had to come clean especially, as by this time I needed two hands to get a drink to my mouth. I decided that I would be able to get through the evening adding that I wouldn't be able to pour drinks. Our guests were very understanding and helpful especially when, after the meal, I found I couldn't shuffle or deal the cards. I could just manage to sort those in my hand. The evening ended when my final bid put us in a slightly dubious slam to be played by my partner. After much thought and considerable tribulation he managed to make the difficult contract, after which he exploded into youthful exuberance quite unbecoming for one of his age.

I then struggled upstairs hanging on to the banister with two hands. Next morning my wife needed to support me to get to the bathroom. I returned to bed while Mary went to Church. On her return it was obvious that my weakness had increased and that further action needed to be taken. It was plain that I needed to be admitted to hospital. By then it was impossible for me to get up let alone go down stairs so it was necessary to ring the emergency hospital service making a "one one one call". This proved frustrating as I first spoke to a lay person who despite my telling her that I was a doctor insisted that I answer all the questions on her chart and tick list. After some twenty minutes she said I would have to speak to the duty nurse and that she would call when free. After a further twenty plus minutes she phoned whereupon, despite being again told I was a doctor and needed hospital admission, I had to answer another set of questions before being informed that my symptoms did not match any condition on her list and that she would have to arrange for the duty GP to visit.

The GP attended within the hour and arrived, having heard about my symptoms from the nurse. From the information he had received, he had formulated a differential diagnosis of three conditions. He was in his early forties, the principal in his practice which took trainee GPs and appeared very much switched on. He took my history and examined me before asking me what I thought I might have. I replied Guillain-Barré Syndrome (GBS). I had ruled out any problems with the brain, such as a stroke, as these almost certainly would not have commenced insidiously and progressed. This condition was very much on his list and, having seen and examined me, was his front runner. He arranged for my hospital admission and a short time later an ambulance arrived. By this time my ability to move and support myself was non-existent and it was necessary for the St. John's ambulance team who came from Taunton some thirty miles away to gently bump me downstairs on a wheeled chair and lift me out and into the vehicle. I arrived shortly after at Dorset County Hospital, Dorchester, and was placed in a side room, while a bed was found for me, avoiding an inevitable delay in the A and E department. Eventually having been examined and assessed by the emergency on call staff and blood tests and x-rays done; I was transferred to the ward. Once there, the Senior Registrar, soon to complete his training and be ready for a consultant appointment, further assessed me. The x-ray of the head and neck had shown a marked abnormality of the cervical spine resulting from an injury when I had severe whiplash some twelve years earlier. For a while it was considered that this might be the cause of my symptoms although GBS remained the front runner.

The spinal unit at Southampton was consulted and the x-rays reviewed over the internet. It was thought that this was unlikely to be the cause of my problems particularly as my paralysis was progressing. The following day a further barrage of tests was undertaken. A lumbar puncture was attempted but I was not surprised when this failed, as it had proved impossible three years earlier to give me a spinal anaesthetic when having a knee replacement. I next underwent an MRI scan.

For this I was placed in a confined tunnel space, not good for anyone who might be claustrophobic, and subjected at various intervals to staccato bouts of sound like a machine gun going off at close quarters. The images obtained gave further information about my brain and the abnormality in the neck. Following this I was taken late in the day to the neuro-physiology department and had an EMG (Electromyogram). While having this it was explained that this test gave a diagnostic trace with GBS. Each time a reading was taken I received an unpleasant electric shock in the area targeted. In GBS the outer covering of a nerve, the myelin sheath, gets destroyed and thus the electrical charges down the nerve are dissipated into the surrounding tissues and fail to reach their target area. The muscles with no electric connection thus rapidly waste away. The EMG report was received the following morning, whereupon I was referred to a consultant neurologist. During this time I was getting progressively weaker and was finding difficulty in coughing. I was unable to move and had to be nursed fully in bed requiring turning, feeding and full help for bodily functions. The neurologist attended and armed with my history examined me carefully, noting the extreme weakness of my muscles and the absence of normal reflexes. The EMG report confirmed the clinical diagnosis of GBS and he wrote me up for treatment with immuno-globulins which only he could prescribe because of the cost.

As GBS began to affect the nerve supply to my diaphragm and my lung function deteriorated, I came under the radar of the ITU and was border line as to whether I needed to be admitted to the unit for observation prior to possibly needing respiratory support. That evening I had the first of five daily intravenous treatments of immuno-globulin and later became aware that the progressively increasing weakness of my body had halted. At this stage I had some weak movement in my left arm, virtually none in my right and minimal in my legs. I could do nothing for myself. I had not the strength to press the patient call buzzer or access my mobile phone. Mary helped me with feeding and I was drinking through a straw. I needed turning two hourly in bed. Not able to move my legs I was a sitting duck for DVT which would have been an embarrassment in the light of my research work many years earlier. I was therefore fitted with TED compression stockings and given a daily subcutaneous injection of an anticoagulant, Clexane. The medical staff had told me that I should recover although it could take up to a year to get back to normal. The days were long as all I could do was to lie in bed, or sit out for an hour or so having been lifted from the bed by a hoist. My body clock had to adjust to the slower tempo which, when achieved, speeded up the passage of time. Visitors helped pass the time as while nearly totally paralysed my brain, if a little slowed, functioned nearly normally. Among my visitors was a doctor friend who had had GBS some thirty years previously. While I appeared to have the acute variety he had had the chronic form with a slower onset. He said it had taken him a year to get back to work and then he was not fully fit. As the treatments progressed, I became aware on a daily basis of a small return of function: the arms where the nerves were shorter and, therefore, required less repair work recovering more rapidly than the legs. Being one with a positive outlook throughout life, I determined to take this on. When I was told that it would probably take a month to achieve a certain function I resolved to set this as a target to beat. Surprisingly with all the support, both physical and spiritual, and encouragement I received I began to achieve my targets rather earlier than those predicted. After two weeks, by which time I was coming off the ITU radar screen, my consultant stated that I was progressing at a pace that he had only seen in one other. This was indeed encouragement.

During my second day in hospital, I was moved from the acute admissions ward to one nearby which was run by a charge nurse who earlier had been in the Royal Navy. As might have been expected there was a military precision to the way routines such as drug rounds were done. I became friendly with him and realized that he was keen to have one of the best wards and nursing staffs in the hospital. I was in a bay of six beds, most of the patients, who had been smokers having been admitted with chest problems and requiring piped oxygen. GBS being such a rare condition (the incidence being about 1: 50,000), patients with this condition were fitted into a ward where a bed was available.

Life in the ward was not without incident and there were occasional scary moments such as when the patient in an adjacent bed died and another opposite collapsed and required full resuscitation to be recovered. About six medical and nursing staff remained round his bed for two plus hours until he was revived and in a stable condition. Because of the low staffing levels none of the ward routines, such as the night drug round, were done until a late hour. Next morning, having thought about the situation overnight, I spoke with the charge nurse pointing out that if another patient in the ward had collapsed at the same time in a similar way he would not have had the slightest chance of survival. The solution, I suggested, might be for the hospital to have a senior nurse on duty doing administrative work overnight who might be redeployed to a ward to give additional cover in a situation like this. This appeared a good example of how the Health Service was managing by flying by the seat of its pants. The charge nurse who also sat on a hospital medical committee raised these concerns at the next meeting. I suppose that while I could scarcely move, the brain with over forty years experience of working in hospitals, was able to function analysing a potentially dangerous situation and divert any thoughts I might have of my own predicament.

As time went by, I became aware that each day I was able to achieve a little more. By two weeks I could use the patient buzzer and could press the control panel to alter my elevation in the mechanical bed. I was now able with difficulty to get food, providing it was cut up, from a plate to my mouth. Rather than having news read to me I was just able, somewhat ponderously, to turn the pages of a paper and hold a pen to make marks soon to become legible and do a sudoku. Physiotherapy was begun initially attempting and encouraging limb movements in bed. The upper limbs recovered more rapidly than the lower but I still was unable to turn myself in bed. Each day I garnered satisfaction and pleasure in being able to do something that had not been possible the day before. I could struggle one day to move a limb shakily and somewhat uncontrollably in a direction and find the next day the movement was much stronger and more purposeful. With these improvements, the physiotherapists in the third week began to encourage me, under supervision, to get out and sit out of bed and to try and stand. On the first few occasions that I tried this, I found the knees wobbly and I collapsed fairly rapidly into a chair. After a few days, I was sat in a wheelchair and transferred to a treatment room. There I marvelled how things had changed over the years. No longer did the therapists have to lift you to a piece of apparatus, one would have a board placed between the chair and couch and you would slide across or to get to a piece of equipment such as parallel bars a hoist would be used. A main feature of GBS, in addition to paralysis, was fatigue and I certainly noticed that the physios were careful to limit how much I did. After a session lasting half an hour I would return to the ward tired and need a period of rest to recover. Another benefit of being able to sit in a wheel chair was that I could be pushed down to the hospital restaurant for a proper cup of coffee and a change of scenery. Among the other patients in my ward was a man who had been a publican in London. He was quite a character who had originally had a pub in Brixton. This he had sold some six weeks before the riots in 1985, when the pub was set on fire, and he had moved to become landlord of another in the Balham area. He flourished and eventually retired to Dorset. He had had GBS some years previously and was able to tell of his experience and give me encouragement daily as I progressed. Considering the rarity of this condition – a small world!

By the fourth week as my medical condition was stable and therefore no longer in need of acute medical services, discussion took place as to where I could be transferred for ongoing rehabilitation and physiotherapy by a therapist familiar with GBS. Enquiries revealed that there was one at the Yeatman hospital in Sherborne which happened to be the nearest hospital to my home and where I had many friends through my activities on the golf course and in the bridge club. I saw the consultant again on the Friday who was once more delighted with my progress. I had been warned that while recovering from GBS I could expect times where my condition could go back rather than forward. I had had one episode lasting five days where this was the case and had been advised that this could continue to occur during my recovery.

During my time at the Dorset County Hospital, my wife had resurrected her nursing skills, very much caring for me during the day, helping to feed me, hydrate me, encourage me and ensure that I was clean and tidy. Her help was much appreciated by the nursing staff who were overstretched and who would otherwise have had to cope with a patient early on needing almost full-time nursing attention. I was a lucky man.

On the Monday, just over four weeks after admission to hospital I was told that I would be moving to Sherborne at around 2.00pm that afternoon. Having seen how other patients had been given times for their transportation only to be delayed for four hours or more, I prepared myself for a wait. Three and a half hours later I was told the ambulance was on the way and would be there shortly. Just before six-thirty it arrived. I left the ward in a wheelchair wishing the other patients well and a speedy recovery, and thanking the nursing staff present for all the help they had given over the previous four weeks. They were skilled and because of shortage extremely hard worked, doing up to thirteen hour shifts with minimal breaks. On top of this, some had journeys home that could take an hour. If they were on the following morning they would leave home at six thirty to arrive for duty and ready to go at seven forty-five.

It transpired that the ambulance had been returning from Salisbury to its base at Gillingham when it was summoned. The crew were happy to be doing this trip as they were on overtime and duly delivered me to the Yeatman around seven. I found it hard to comprehend how the service could be so far out on its timing particularly as there were two groups of ambulance, those available for emergencies and those for transportation trips such as my journey from Dorchester to Sherborne.

I was put in a four bedded ward. At this point I needed to be moved around in a chair but could turn in bed by pulling on cot sides. The equipment in the way of hoists was amazing, most of which had been purchased by the League of Friends. Rather than having a bed bath, as had been necessary at the DCH, I could now have a proper bath. I was placed on a hoist and lifted over the side and into a bath. Not only was the hoist mobile but the bath could also be elevated. What luxury after more than a month. The therapists took me into the treatment room adjacent to the ward and assessed my capabilities and wrote a program of daily exercises including the number of repetitions to be done. Standing and stepping on to a block was almost impossible at first, and I was made to understand that doing exercises quickly was disadvantageous, benefit accruing only when they were slow and controlled. Among apparatus used were parallel bars and a cycling machine which could be set and graduated daily to give an increase in difficulty and so build up my leg muscles. Using this not only did the legs tire but unaccustomed to exercise as I was, I became breathless and needed a rest and rehydration. On trying to stand, between the parallel bars on the first occasion, I found I had no control of the left leg and that it was wobbly and collapsing. Also my sense and ability to balance was impaired. By now I fully appreciated that what I couldn't do today, providing I was sensible and followed instructions, would be possible in a few days time.

At the Yeatman, I had a host of visitors, those from the golf club kindly being organized by a friend so that I did not receive more than two on any occasion. I valued these visits enormously as they gave an opportunity to discuss current affairs, in particular the upcoming European referendum and enabled me to focus on events outside the hospital. This was a major relief as the other patients in my ward could not be sociable. One had given up the fight and was slowly dying, while another, a Scotsman, was demented and continually rambling about the past. One of my duties became keeping an eye on him as he would intermittently try to get out of bed on to legs incapable of supporting him. This came to be the major demand on my patient buzzer. The third occupant was pleasant and obviously had a good sense of humour but was markedly deaf which made communication other than by nearly shouting difficult. As the days passed my wife, in view of all the visitors I was receiving, was able to take time out and return to some engagements she had been unable to keep. Not only was I receiving local visitors but my daughters in the UK and my brothers and their wives all made multiple trips to spur me on. Having read about GBS on the internet they were surprised by my rate of progress.

Once my treatment regimes were established, I made steady progress building my arm strength by propelling myself in a wheelchair and beginning to walk, at first, with a frame. To begin with, while walking, I needed an escort to support me should my legs collapse or I lose my balance. Over a week I progressed to starting to walk with crutches before advancing further, as my strength and balance improved, to managing with sticks. I quickly learned that what was difficult or nearly impossible one day was almost invariably accomplished a day or two later. This was indeed good for morale. With my recovery fortunately speeding along, arrangements were made for a home visit with an occupational therapist to assess what aids I might need on my return and to see if I could manage the stairs and the home environment. This visit was made two and a half weeks after my transfer to the Yeatman and more than six weeks after my admission to hospital. Both my ambulance trips had been in the dark, so what joy it was to see the countryside once more and to arrive home. My assessment was accomplished without problems and it was agreed that, together with a number of support props that I might possibly require, I could return home the following week. In view of my speed of recovery, I felt I was being over provided for, but nevertheless accepted them in case I found back up and assistance was needed.

Back in the Yeatman, I resolved to advance my recovery as far as possible during the remainder of my stay. I pushed the exercises as far as allowed but was careful not to go beyond what the therapists said I should do. I did not want to have an accident because of my own stupidity and end up with further time in hospital. Without mishap and having made considerable further progress, I was discharged from hospital seven and a half weeks after first being admitted.

I arrived home and, as I had suspected, had made considerable progress since my visit the previous week. I was able to get around with two sticks but took particular care going both up and down the stairs. I continued to make good progress and was soon walking around the house with a single stick. I began to venture outside and daily increased the distance I could walk. As my recovery continued at a good pace with my wife ensuring that I continued with my exercises, I began to muse on the previous two months.

I had gone from playing golf one day to being to all intents totally paralysed within forty-eight hours. I was fortunate in that I had been admitted to hospital rapidly and that the diagnosis was arrived at quickly, so that treatment could be instituted without delay, thus saving me an otherwise almost certain admission to the ITU and respiratory support. Only later did I contemplate what might have happened if I had been in a remote area. Not only was it a shock for me but more so for my wife. She was fortunate in that a friend, a retired doctor who had himself had GBS, was able to talk her through it and put her in touch with GAIN a charitable body supporting patients with this condition. They sent literature about GBS covering it from A to Z which proved a great help. She received many many phone calls concerning my state and offering support and was able to direct them to Google for information and an explanation of the condition.

I was fortunate to have excellent nursing care and physiotherapy and support both spiritual and physical from my family and many friends which helped to speed me, while in hospital, on the road to recovery. It was an additional boost to learn from the consultant, physiotherapist and nursing staff that, considering the severity of my condition prior to treatment, I was making one of the fastest recoveries that they had seen.

GBS is an experience I would not wish on anyone. It was frightening at first, both for me but also my wife, until I underwent treatment and began to respond. It then became a challenge to overcome in order to get back to a normal way of life. It helped having a medical background and realizing that it would almost certainly be only a matter of time, the forecast being up to a year before normal service could be resumed. I was warned that it was likely I would need to be in hospital for three to four months undergoing treatment and rehabilitation. I was lucky in that with all the support I had, this prognostication was considerably shortened.

Once home, I continued with the exercises I had been taught. I slowly increased the distance I was walking and once a week attended the physiotherapist at the Yeatman Hospital to further, under supervision, advance my recovery. One day, when out walking in the village I bumped into a recently retired Physician who wished to know what I had been through. After recounting the lurid details of my illness and seeing where I then was, as we separated he commented “Tough buggers, you surgeons”.

Once discharged by the physiotherapist I was referred to the Sherborne Sports Centre where under the supervision of a trainer I was given balancing, strengthening and stamina increasing exercises, which I did twice a week. I continued to make rapid progress. Soon I was attending the driving range at the golf club, again moving forward gradually as my stamina increased. After first playing nine holes using a buggy as I did prior to my illness, I managed to play eighteen holes just four months after the onset of the GBS. I had been incredibly fortunate to get to this stage in this time, helped and spurred on by all the support I had had from the medical and physiotherapy staffs from family, friends and spiritually, which confounded the prognostications made at onset regarding my recovery.

Having worked in hospitals for forty years I can now say truly I have seen life from the other side of the fence. I really appreciated the skills and dedication of the nursing staff, in the wards I was in, and the demands made upon them, often working under extreme conditions due to shortage of personnel and long hours. It was also an experience being in a confined space with other patients and realizing that you were the fortunate one who could hope and expect to return to a normal life, an option not possible for most of the others.

Chapter 18

Some Final Reminiscences

I have been lucky in life, first to have parents who were both surgeons thus giving me a chance to inherit the correct genes and, secondly, being brought up in a hospital environment and learning at an early age what career I wanted to follow. I could focus on my future and enjoy the opportunities available while progressing through youth without worrying about what might lie ahead. I was fortunate in that, at that time, surgical staff were housed on site in properties owned by the hospital and with recreational facilities present. The flat roof of the treatment centre gave me the chance to view operations as from a gallery and, for safety reasons, to be invited into the theatre and see them close at hand. In retrospect it was a privilege to be able to scrub up and assist at operations from the age of fourteen and learn to tie surgical knots, insert skin sutures and to be guided through my first operation at sixteen. With a start like this, it was inevitable once the basic exams were completed that I would be looking to take the next step forward. Senior surgical staff having viewed my capabilities and realizing that I would not overreach but summon them when up against the unknown or in difficulty allowed me great latitude and so gain experience, particularly in the numbers and variety of operations performed.

I learned at an early age that while performing an operation for example on acute appendicitis or a strangulated hernia, no two operations for either of these conditions needed to be the same; one could be utterly straight forward while the next at the other extreme being complicated and most difficult. Some patients would be thin while others obese, some patients could have early pathology while others advanced. Man was not a machine with every model being the same but, in extreme, subject to much variety ranging from congenital abnormality to opposites of pathology, some being early some late as with inflammation or the presentation of cancer. The response of patients to surgery could also be poles apart, most taking procedures in their stride while sometimes the more elderly or those with other pathologies, such as diabetes or respiratory problems proving more of a challenge to the surgeon at operation or during subsequent recovery. While some that you would expect to progress without problem did not, others in whom complications were likely made uneventful recoveries. With so much variety the carer could take nothing for granted.

After being qualified for two years, I thought my surgical career was set in stone. Little did I visualize the enormous changes that were to occur. The advances in anaesthesia, with rapid post-operative recovery, enabled patients to be considered for operation irrespective of their age. Pathologies such as peptic ulcer, with the arrival of curative treatment, virtually disappeared from the surgical scene. Advances in technology producing heart bypass machines allowed open heart surgery to be done while, with the ability to negate rejection with immuno-suppression, transplantation became successful. In the late nineteen-eighties keyhole surgery in various forms such as orthopaedic or laparoscopic was developed. Nothing is standing still, with robot assisted keyhole surgery allowing speedier and quicker recovery from some forms of heart surgery and kidney transplants, with probably other forms of treatment in development. It is an ever changing world and for the benefit of the patient it behoves the surgeon to face the challenges and keep pace.

Before setting out, having had no previous experience, I was sceptical about doing research believing it to be a necessary but not a very rewarding step to take on the ladder to achieving in a desirable place consultant status. It was in this frame of mind that I set out to face the challenge put to me by Frank Cockett and certainly would have remained so had I been the next cog in the wheel of an ongoing research project as was in progress on the surgical professorial unit. It was humbling to read papers written by highly respected and talented surgeons and researchers throughout one hundred years and to believe one could make any further contribution unless I could find some form of technology, not previously available, to enable more precise examination of the peripheral venous system.

Hopefully this could give a passport to examine the peripheral circulatory system as never before.

As previously related, Doppler ultrasound capable of examining venous blood flow from behind the knee to the upper abdomen became available. Studies which I performed, rapidly demonstrated its efficacy, showing a ninety-five percent correlation with the gold standard but somewhat time-consuming radiological studies (peripheral phlebography). When comparing this figure with the accuracy of clinical examination when only fifty percent of those thought to have a thrombosis indeed had one and in a further twenty percent the diagnosis was overlooked, Doppler ultrasound proved a major diagnostic advance. Today, incorporating further technological advances, it is the primary form of investigation with clinical, screening and research applications. The correlation between ultrasound and the gold standard examination has remained at ninety-five percent till this day. A clinician today can take a Doppler Duplex ultrasound scanner out of his pocket on a ward round and in less than five minutes examine a patient there and then.

When I first became aware of hospitals the NHS was not in existence. The management of Lord Mayor Treloar Hospital was in the hands of the Medical Superintendent, in this case my father, a hospital matron, a hospital secretary and a steward, the latter being responsible for the non medical day-to-day requirements. This small committee was overseen by a board of Trustees. In my father's case at Alton among the Trustees was a leading London lawyer, a London banker, an editor of a daily national newspaper, the daughter of the founder, Miss Florence Treloar, and others of similar standing. Following nationalization these eminent members were replaced in the course of time by local dignitaries and shop owners who in no way had the experience in managing budgets of this size. In 1948 the hospital became part of the NHS.

Over the years, hospitals came more under the control of CEOs and administrative staff, with medical staff as time passed having less and less influence. Until the 1990s this seemed to have relatively little effect on patient care and management, although there were cut backs in the number of hospital beds and staff. Throughout this period though there was a steady increase in administration. In the

1990s, the Government decreed that waiting lists for hospital admission and in particular surgery should be no more than two years. This led to a mushrooming of administrative staff to ensure this was implemented. Patients on waiting lists were normally categorized by the surgeons as Urgent, Soon or Routine. Those on the Urgent list usually had life threatening conditions such as cancer or an aneurysm which could burst at any time. On the Soon list were patients having serious problems with gall stones, hernias threatening strangulation and others with pathology seriously interfering with their everyday life. Those awaiting Routine admission had problems not really affecting their day-to-day existence but where corrective surgery at some stage was advisable or others wishing removal of lesions often for cosmetic appearances.

It appeared to the clinician that hospitals were moving towards being run as a business where there appeared to be little concern for the patient or the patients' pathology or symptoms. An exception to this was the treatment of cancer, an emotive issue but one where early diagnosis and therapy could make for a good outlook. At the time of my retiral attempts were being made to implement the two year admission strategy. It was my strong belief that doctors should continue to manage patient admissions and care and should not tolerate interference from administrative staff that is in no way qualified to take this responsibility. The administration should supply the back-up to ensure the smooth running of the facilities necessary for quality patient care and the presence of sufficient staff numbers to offer proper and safe treatment. Hospital matrons had virtually disappeared, this again being to the disadvantage of patients' well-being, as there was little check on nursing standards and maintenance of ward cleanliness.

While the NHS was free at the point of delivery and this had become a sacred cow for all political parties, the patients were nevertheless being let down by the standard of treatment they were receiving in relation to other countries throughout Europe and the World. In Europe the overall results of treatment in the UK, despite centres of excellence, for a variety of conditions were poor. Nowhere in comparative lists published was the UK in the top ten out of twenty-eight of results achieved and mostly resided in the early twenties. This demeans and demoralizes the teams of carers who know that if properly organized their results worldwide should be in the top percentile and at the same time is an injustice and a let down to their patients. I find it very sad that people as a whole focus on the access to the health service rather than the results obtained. The political emphasis sadly is focused on the wrong place. Rather than considering access, it should be looking at outcomes and enlighten the population accordingly. I feel certain that if attention was focused on this and the public understood that they were not overall in the premier league for results, there being a postcode lottery, then they would be willing to accept some different form of funding as exists in many countries throughout Europe and the Commonwealth. This would enable there to be more medical staff, more rapid access to a doctor and the best treatment with, for example, expensive drugs to allow for enhanced outcomes. Without rationing and with speedier access to ever improving diagnostic facilities and treatment, the UK generally would move up the league table and become once more a leader in patient care, management, and outcomes achieved.

As the number of beds in hospitals was reduced, it was fortunate that keyhole surgery arrived as this very much changed the speed of post-operative recovery, allowing patients to return home in a third to half the time that was standard with open operations. Keyhole surgery thus enabled the same throughput as had previously been possible. However because of the cutbacks, cancellation of operating lists became common in winter as the number of medical admissions due to 'flu epidemics and associated lung infections caused an overflow into surgical beds. Around this crunch time operating lists per se needed cancelling, giving the surgical and anaesthetic staff time to follow other pursuits.

In the last quarter of my surgical career with the arrival of keyhole surgery and seeing the potential benefits to the patients, I was inspired to take this on. This was somewhat against the flow of opinion from others. Some thought I was mad to tackle this on the premise that you can't teach an old dog new tricks; whilst others believed it would be here today and gone tomorrow. It was a shot in the dark and a little presumptuous to believe that firstly I could raise the finance to buy the equipment and fund my training and secondly that I would have the confidence and ability to undertake new challenges and acquire new skills. If successful, I could see the benefit for the patients, while if not there was always the option to revert back to open operations but at a cost for failure of the original outlay of one hundred thousand pounds. I tried to raise money in the West Midland region, but with no success. However luck was on my side in the form of the chair of the League of Friends who proved a fairy godmother (June Whitaker). After reporting to her what I had learned in London and how I had performed on a simulator, she invited me to speak to her committee and then backed me up to the hilt, persuading its members to endorse her wishes. Some years later and now retired, my "fairy godmother" who had also retired, said on meeting that during her many years in association with the League of Friends, the setting up of keyhole surgery had been her finest achievement.

Finally, I reiterate that I have been lucky in life. I have a wife who has supported me throughout all these challenges. I have been fortunate to have been born with the necessary latent skills to take on keyhole surgery. It was fortuitous that during my time as a researcher, technology had advanced sufficiently to enable the bedside diagnosis of deep vein thrombosis to be made and evaluated. During my training, I had been fortunate to work with some of the finest surgeons in the country, from whom I learned a great deal. I had been brought up in a hospital environment and much had rubbed off on me during those years. In retirement I have found new challenges and I have no doubt as I approach the twilight of my life there will be still more. However, I was equipped with the correct genes from my parents to enable me to say I was born to be a surgeon.

Printed in Poland
by Amazon Fulfillment
Poland Sp. z o.o., Wrocław